# Contemporary Perspectives in Data Mining

## Volume 2

A volume in
*Contemporary Perspectives in Data Mining*
Kenneth D. Lawrence and Ronald Klimberg, *Series Editors*

# Contemporary Perspectives in Data Mining

## Volume 2


*edited by*

**Kenneth D. Lawrence**

*New Jersey Institute of Technology*

**Ronald K. Klimberg**

*Saint Joseph's University*

INFORMATION AGE PUBLISHING, INC.

Charlotte, NC • www.infoagepub.com

**Library of Congress Cataloging-in-Publication Data**

A CIP record for this book is available from the Library of Congress
http://www.loc.gov

ISBN: 978-1-68123-087-0 (Paperback)
978-1-68123-088-7 (Hardcover)
978-1-68123-089-4 (ebook)



Printed in the United States of America

# CONTENTS

## SECTION I

### MARKETING APPLICATIONS



## SECTION II

### BUSINESS APPLICATIONS

## SECTION III

### TECHNIQUES

# SECTION I

MARKETING APPLICATIONS

CHAPTER 1

# DATA PRIVACY IN LOYALTY PROGRAMS

## An Exploratory Investigation

**David Burns and Gregory Smith**
***Xavier University***

### ABSTRACT

The right to personal privacy has been and remains a hot-button issue in modern society. This tenet is at the heart of the American Dream for many. However, modern technologies and methods have awakened some to a new area of concern: personal data privacy. Recent headlines have revealed that both private and public institutions actively collect and monitor personal data. As consumer citizens, the most obvious and identifiable area is the use of retail loyalty cards and their associated programs. These programs allow retailers to have access to the "who, what, when, and where" of a consumer purchase. Unfortunately, little research has been done to capture the sentiment that consumers have towards their privacy, loyalty, and the loyalty card programs they use. The following chapter provides an introductory discussion, insights, and the results of a broad-based sentiment survey on these topics. The results of the survey are then used to test several longstanding hypotheses about privacy and loyalty.

*Contemporary Perspectives in Data Mining, Volume 2*, pages 3–23

## DATA PRIVACY IN LOYALTY PROGRAMS: AN EXPLORATORY INVESTIGATION

In the United States, personal privacy has garnered attention since the nation's founding. Recently, advances in technology changed the nature of the privacy discussion. However, technology, more specifically data mining, in and of itself does not threaten privacy. The threats stem from how technologies are used and affect social life (Wacks, 2010). Concerns regarding possible threats to privacy emanating from advances in technology are not new (Bélanger & Crossler, 2011). Over a century ago, for instance, Warren and Brandeis stated that "instantaneous photographs and newspaper enterprise have invaded the sacred precincts of the private and domestic life; and numerous mechanical devises threaten to make good the prediction that 'what is whispered in the closet shall be proclaimed from the house-tops'" (1890, p. 195). Warren and Brandeis spoke of advances in photography and in printing technology that permitted the widespread dissemination of photographs taken of individuals without their permission. Today, the concerns raised by Warren and Brandeis seem trite. Since the time of Warren and Brandeis, a seemingly endless array of new technologies has been introduced that have the ability to be used to directly impact individuals' privacy (Smith, 2000). Advances in technology have made surveillance much easier and much more pervasive, allowing for both direct and indirect monitoring. Information can be gathered and data mined in a vast array of methods, often without the knowledge of the individuals whose information is being gathered.

Since the activities of retailers focus on satisfying their customers, most have gathered and maintained information on these customers to use in ways to increase patronage through the use of data mining. Consequently, retailers have consistently been on the on the cutting edge of identifying and utilizing specific technologies that improve information-gathering and -processing abilities. Today, a primary method by which personal customer information can be acquired, maintained, and eventually shared involves the use of loyalty cards, which is the focus of this study. Loyalty cards have proven to be a very successful means to acquire personal information, so much so that significant privacy concerns have been raised (e.g., Davenport & Harris, 2007). The objective of the study is to provide insight into the relationships between consumers' desire for privacy and their opinions about loyalty cards and their use. First, we explore the concept of privacy. Second, we examine customer loyalty. Third, loyalty cards are examined and discussed. Fourth, we develop and test several hypotheses pertaining to privacy. Finally, conclusions are drawn.

## PRIVACY

Although privacy is a commonly used term, agreement is lacking on a common definition (Wacks, 2010). Ordinarily, privacy is characterized either as a limitation on access to or as a form of control over one's own information. Research most frequently characterizes privacy as a form of control that one possesses over one's information. Westin, for instance, defines privacy as "the claim of individuals, groups, or institutions to determine for themselves when, how, and to what extent information about them is communicated to others" (1967, p. 7). Nevertheless, the research that views privacy from an access perspective tends to be more developed. As quoted in Nissenbaum (2010), Gavison (1980) defines privacy from an access perspective as "a measure of the access that others have to you through information, attention, and physical proximity" (p. 68). Allen-Castellitto has combined these perspectives on privacy by defining privacy around three dimensions: physical privacy, or "special seclusions and solitude"; informational privacy, or "confidentiality, secrecy, data protection, and control over personal information"; and proprietary privacy, or "control over names, likenesses and repositories of personal information" (1999, p. 723).

Van den Hoven (2007) makes a case as to why privacy deserves protection. Regardless of the technology, the issues of concern appear to be (1) the ability monitor and track individuals over time and space, (2) the capacity to store and analyze these data including the ability to merge data from a number of diverse sources to build individual profiles, and (3) the dissemination and publication of this data, which includes its selling (Nissenbaum, 2010). Recent revelations about the data-gathering actions by the U.S. government through its PRISM program and the construction of the Utah data center have brought renewed attention to the issue of privacy.

Technology permits the acquisition of vast amounts of data, often without the individuals' awareness. Due to video cameras, emails, phone calls, Internet searches, television viewing habits, retail loyalty programs, and a host of other technologies, privacy is arguably something that has vanished from the lives of most individuals. Consequently, research appears to indicate that privacy concerns are important to many. A 2003 Harris poll, for instance, indicated that 69% of respondents agreed that consumers have lost control over how their personal information is collected and mined (Laczniak & Murphy, 2006) and a Pew Internet Project found that 85% of individuals believe that it is "very important" to control access to personal information (Madden, Fox, Smith, & Vitak, 2007). Some individuals predict that privacy rights will be to the 21st century as civil rights were to the 20th (Laczniak & Murphy, 2006).

Research suggests that consumers with differing backgrounds possess varying levels of concern for privacy. Milne and Bahl (2010), for instance, observed that males and individuals with higher education and higher income tend to possess less concern for privacy than females and those with less education and lower income. O'Neil (2001) also observed that females and individuals with higher income possess higher privacy concerns. Likewise, O'Neil (2001) also observed similar results for education (higher educated individuals possess higher privacy concerns).

It is logical to expect that individuals' concerns for privacy affect their choices and activities, including those in the marketplace. Although it has not been empirically examined, it would seem that individuals with greater concerns for privacy will pursue actions that would seemingly limit the dissemination of their personal information to others, such as doing business primarily with brands and stores that they trust. Such deterministic actions would lead to brand and store loyalty.

## LOYALTY

Customer loyalty is highly prized by virtually all businesses. Customer loyalty provides businesses with repeat customers, providing a continuing stream of business. Without repeat customers, businesses would be faced with the need to continually attract new customers. Given the costs attracting new customers (often running several times the costs of retaining existing customers) (Omar, Alam, Aziz, & Nazri, 2011), building customer loyalty is typically a very important component of the marketing strategy of most businesses. Bellizzi and Bristol (2004) note that customer loyalty is particularly important for retailers. Indeed, the results of a recent study sponsored by the National Retail Federation (NRF) indicate that retailers believe retaining customers to be their top corporate strategic priority ("Survey: Satisfying and Retaining Customers Is Top Priority" 2009). This is not surprising. Consumers are often faced with a large number of retailers from which to choose. Hence, retaining customers appears to be an area that warrants particular attention.

Loyalty can be expected to connect with the level of trust a consumer has with a brand or a retailer. Trust is typically viewed as essential for building and maintaining successful relationships (Berry, 1995) and is included in most relationship models. Zeher, Şahin, Kitapçi, and Özşahin (2011) view loyalty as resulting from an ongoing process of continuing and maintaining a relationship created by trust. Consequently, trust is typically viewed as essential in building long-term loyal relationships with customers (Peppers & Rogers, 2004).

Empirical relationships have been observed between trust in a store and loyalty to that retailer (Mohammad, 2012; Omar et al., 2011)—individuals tend to repeatedly patronize retailers in whom they possess trust. Similar relationships have been seen between trust in a brand and loyalty to that brand. The research indicates that brand loyalty is an important cause of long-term loyalty and that it strengthens the relation between brands and consumers (Geçti & Zengin, 2013; Liu, Guo, & Lee, 2011). Trust in a brand results from a considered process (Chaudhuri & Holbrook, 2001) and has been shown to directly affect brand loyalty (Halim, 2006; Sung & Kim, 2010). Brand trust can also be viewed as a lack of trust in other brands. This lack of trust in other brands can further build loyalty to a brand (Ibáñez, Hartmann, & Calvo, 2006).

Loyalty can lead to situations where trust can be enhanced. Loyalty can permit a business to become more informed about the wants, needs, and nature of its loyal customers, allowing them to be in positions to better serve them. For very small businesses with few customers, this activity can be accomplished informally as one-on-one relationships can be built with each customer. Larger businesses, however, face a more difficult situation. As one-on-one relationships with each customer become impossible, larger businesses are forced to rely on alternative means to understand their customers. Today, several large businesses use technology to aid in building and maintaining customer loyalty, usually through the process of building database systems, allowing them to identify customers and track customer transactions. Such customer databases permit businesses to better address customers' wants and needs, and to be able to predict changes in customer purchase patterns (Batislam, Denizel, & Filiztekin, 2007). One of the most common ways that such customer databases can be developed in retailing is through the use of loyalty cards.

## LOYALTY CARDS

The popularity of loyalty cards is undeniable. Loyalty programs are one of the most popular strategies used by retailers to increase store loyalty in mature retail markets (Omar et al., 2011). Today, over 27 million retailer loyalty cards are in circulation in the U.S. (Aitken, 2007). In Canada, loyalty cards diffuse at a greater rate, with 97% of Canadians possessing the loyalty card of at least one retailer (Robson, 2006). Loyalty cards are a relatively recent phenomenon, beginning in the 1990s primarily as a means to develop and maintain store loyalty. Loyalty cards provide customers with inducements (or rewards) for using the card, either in the form of cost savings available through mass promotions (such as limiting most sale pricing to card holders) or through the use of cost savings through customized deals based on specific customer's demographics or shopping behavior (Pauler & Dick, 2006; Smith, 2008). The goal of the rewards is to increase customers' allegiance to the retailer and to increase the "share of wallet" spent by customers at the store (Allaway, Gooner, Berkowitz, & Davis, 2006; Cortinas, Elorz, & Mugica, 2008). Consequently, some view loyalty cards as "bribery cards" since they were viewed as a tool to attempt to "buy" customers' loyalty.

## Store Loyalty

Store loyalty is comprised of two components: affective loyalty (attitude toward the store) and behavioral loyalty (purchases at the store) (Dick & Basu, 1994). The behavioral component is regarded as being dependent on the affective component—loyalty exists when customers have a favorable attitude toward the store which leads to repeat purchasing behavior (Demoulin & Zidda, 2008). The behavioral dimension is the usual focus of research attention since it represents actual behavior.

How successful are loyalty cards at building behavioral store loyalty? The widespread adoption of loyalty cards by a significant number of retailers would tend to signify that loyalty cards are indeed successful at increasing consumer loyalty. Surprisingly, however, this does not seem to be the case (Capizzi & Ferguson, 2005). In fact, there is relatively little evidence that retail loyalty cards successfully increase store loyalty (Magi, 2003; Mauri, 2003). Smith, Sparks, Hart, and Tzokas (2003), for instance, did not observe a significant difference in the shopping behavior between members of retail loyalty programs and nonmembers. Similarly, Bellizzi and Bristol (2004) did not observe a relationship between the adoption of a loyalty card and subsequent store loyalty, and Allaway et al. (2006) report that few consumers possessing loyalty cards indicate that they would give up patronizing other retailers to focus their purchasing at the card-issuing store.

The counterintuitive results likely stem from a competitive effect—once loyalty cards are introduced by a retailer in a specific market, often many of the retailer's competitors introduce their own loyalty programs in response (Demoulin & Zidda, 2008). The result is "card saturation," where the existence of multiple programs negates the effects of the initial program (Sharp & Sharp, 1997). Evidence seems to suggest that this is occurring in many markets. Magi (2003), for instance, reports that loyalty cards are most effective when customers do not possess the cards of competitors. Such a situation is not common, however. Most consumers possess loyalty cards from multiple competitive retailers. When consumers possess several cards, the expected loyalty effects from any individual card appear to be confounded out with minimal resulting changes in shopping behavior (Demoulin & Zidda 2008).

Even in the midst of competition, Demoulin and Zidda (2008) note that retailers appear to be able to affect the effectiveness of their loyalty programs. Specifically, they observed that consumers who are more satisfied with the benefits from a loyalty program are more likely to patronize the store who issued the card to a greater extent than consumers who are less satisfied. Furthermore, they are less likely to hold the loyalty cards of competitors. Consequently, Demoulin and Zidda (2008) suggest that loyalty cards can play a successful competitive role. Given that the rewards available from a loyalty program can easily be replicated by competitors, however, Demoulin and Zidda's assessment may be overly optimistic.

Not to be overlooked, however, is that loyalty cards may be able to indirectly affect consumer loyalty. The consumer databases that retailers can develop through loyalty programs can provide retailers with insights into their customers not available through other means (Palmer, McMahon-Beattie, & Beggs, 2000). These additional insights, in turn, can be used to increase customer satisfaction. Indeed, the primary value in loyalty cards appears to be their indirect effect on customer loyalty by providing data to use to optimize marketing strategy (e.g., Felgate, Fearne, Di Falco, & Garcia Martinez, 2012; Passingham, 1998; Ziliani & Bellini, 2004). Category management, for instance, can be facilitated by the data from loyalty cards by permitting retailers to "improve their inventory policy, distribute product space more efficiently, and design a better pricing and promotion policy" (Cortinas et al., 2008, pp. 60–61).

The data derived from loyalty cards can also comprise a significant revenue source in itself (Burns & Toncar, 2011). Many retailers have found that they can sell the data collected via their loyalty cards to vendors and other information-gathering companies for substantial amounts of money. As would be expected, the selling of personal data has raised significant privacy concerns among many.

## Privacy

Advances in tracking and monitoring of individuals' purchases through loyalty cards would be of relatively little concern if the data could not be retained or if it could not be shared with other organizations. Although some questions over the amount of data stored in large stand-alone databases have existed since the 1960s, today, the concern is much different since technology has revolutionized how data is stored and used (Lohr, 2010; Nissenbaum, 2010). With the rapid declines in the price of data storage and the ease by which data can be stored, vast amounts of data on individuals to be retained indefinitely with minimal cost or effort. The data provided by loyalty cards, for instance, can easily be retained and used to affect the shopping environment to the advantage of the retailer that gathered it. More importantly, data can be easily transferred, sold, and combined with other data sets. Data records of individual activities can be simply and elegantly aggregated—data from such disparate sources as a retailer's loyalty program, government records, retail purchase records from other retailers, health records, legal history, financial history, online search activity, social networks, and so on can easily be combined to form large aggregated databases.

Data from loyalty cards can be aggregated with data from other sources to form in-depth profiles of the lives of individuals to better understand individual consumers or to identify consumers as members of homogeneous groups or categories (Payne & Trumbach, 2009; Schermer, 2011). Individual profiles can be very advantageous to businesses (Klabjan & Pei, 2011). The data included in individual profiles can be used to uncover predispositions and to predict future behavior and choices (Nissenbaum, 2010) through the process of data mining. Aggregating information on individuals is not new. Small store owners, for instance, would watch for information in their local newspapers that may provide information to better understand their customers. "But aggregation's power and scope are different in the information age; the data gathered about people is significantly more extensive, the process of combining is much easier, and the computer technologies to analyze it are more sophisticated and powerful" (Solove, 2006, p. 506). Du, Kamakura, and Mela (2007) highlight the value of a business acquiring customer data from their direct competitors. Furthermore,

> A supermarket's database could be used by divorce attorneys in child-custody cases to discredit spouses who have histories of that include tobacco or alcohol. In an actual case, the Drug Enforcement Administration (DEA) subpoenaed a grocer for the purchase records of a customer suspected of drug dealing. (Bellizzi & Bristol, 2004, p. 145)
>
> If records of people's food shopping and therefore eating habits are created and maintained, there is a potential for such data to be used by health insurance companies to raise premiums. (Aitken, 2007, p. 12)

Davenport and Harris (2007) suggest a number of additional scenarios where a customers' data can be used by businesses.

Building consumer databases provides retailers with the means to data-mine, or to search for behavioral patterns in the data that can be identified (Payne & Trumbach, 2009), providing considerable insight into individual consumers. This ability was clearly displayed when Target was able to determine an individual's pregnancy before her family members were aware (Duhigg, 2012). Regarding the use of data form loyalty cards, Payne and Trumbach state:

> The consumer loses aspects of privacy as all of their basic demographic information, personal interests, correspondence and activities are stored in databases and available to be combined together. With so much data available on the consumer, consumers have a lack of control over what happens to that data. Laws do not clearly define the restrictions that companies have based on what they say in their privacy policies. Consumers also face a more sinister

> problem, namely potential discrimination based on the information they provide or refuse to provide. (2009, p. 243)

Furthermore, Laczniak and Murphy state:

> Disturbingly, these profiles are then copied and sold to other marketers who use it to predict likely purchase prospects for their goods and services. As a result, a growing and permanent record exists of what individual consumers buy, where they bought it, the price paid and the incentives that motivated the transaction. (2006, p. 315)

Although individuals have little control over much of their information, consumers can exercise a measure of control over access to their purchasing information by affecting their purchasing locations and whether they join loyalty programs.

## THE STUDY

The purpose of the study is to provide insight into respondents' views toward privacy and the outcomes of those views. First, based on the above discussion, it appears logical that there is a relationship between one's views toward privacy and the loyalty exhibited toward brands and retailers. Loyalty with a brand or with a store is affected by the level of trust that a consumer has toward a particular brand or with a particular store and/or the level of distrust with other brands and retailers. Individuals who are concerned about privacy can be expected to concentrate their purchases on brands and retailers that they trust; hence, they are likely to exhibit higher levels of loyalty to brands and retailers than individuals who are less concerned about privacy.

> **H1:** *Individuals who are more concerned about privacy exhibit higher levels of loyalty to brands and to retailers than individuals who are less concerned about privacy.*

Participating in a loyalty card program involves sharing a significant amount of information with a retailer. It is logical to expect that individuals who are more concerned about privacy are less likely to use loyalty cards, more likely to realize how the data collected by a loyalty card may be used (less likely to view loyalty cards as merely "coupon cards"), and more apt to read the user agreements of loyalty cards.

> **H2:** *Individuals who are more concerned about privacy are less likely to use them for everyday purchases, such as food, are less likely to view loyalty cards*

*merely as "coupon cards," and are more likely to read the user agreements of retailer loyalty cards than individuals who are less concerned about privacy.*

Individuals' views toward privacy are likely related to their level of trust in how various institutions will use their information. Individuals who are more concerned about privacy can be expected to possess less trust in the institutions who use their information. These feelings of lack of trust can be expected to extend to other areas, such as trust in the media.

**H3:** *Individuals who are more concerned about privacy possess less trust in the media than individuals who are less concerned about privacy.*

Although not subject to an extensive amount of research, existing research suggests that individuals' views toward privacy may vary across different consumer demographics. As discussed above, the research seems to indicate that males and individuals with higher income possess less privacy concerns than females and individuals with lower incomes. The research on the effects of education has produced mixed results, but it is logical to expect that the relationship should be similar to the relationship with income. Lastly, given that older consumers tend to more wary of new products and new technologies, it is logical to expect that they possess greater privacy concerns than younger consumers.

**H4:** *Individuals who are more concerned about privacy possess different demographics than individuals who are less concerned about privacy. Specifically, individuals who are more concerned about privacy are more likely to be:*

- *a. female*
- *b. older*
- *c. less educated*
- *d. earn less income*

*than individuals who are less concerned about privacy.*

## METHODOLOGY

### Sample

The sample was comprised of 1,110 members of a large consumer panel in the U.S. The sample appears to be representative of the general population—51% of the sample is female with 72% Caucasian, 12% African-American, and 12% Hispanic. Sample demographics are displayed in Table 1.1.

**TABLE 1.1 Sample Demographics**

| | Frequency | Percentage |
|---|---|---|
| **Gender** | | |
| Male | 540 | 48.6 |
| Female | 567 | 51.1 |
| Not provided | 3 | 0.3 |
| **Ethnicity** | | |
| Caucasian | 804 | 72.4 |
| African American | 138 | 12.4 |
| Hispanic | 133 | 12.0 |
| Asian/Pacific Islander | 11 | 1.0 |
| Other | 24 | 2.2 |
| **Age** | | |
| 18–24 | 116 | 10.5 |
| 25–34 | 240 | 21.6 |
| 35–44 | 236 | 21.3 |
| 45–54 | 193 | 17.4 |
| 55–64 | 136 | 12.3 |
| 65 and above | 189 | 17.0 |
| **Education** | | |
| Some High School | 33 | 3.0 |
| High School Graduate | 232 | 20.9 |
| Some College | 361 | 32.5 |
| College Degree | 301 | 27.1 |
| Some Graduate Study | 59 | 5.3 |
| Graduate Degree | 124 | 11.2 |
| **Income** | | |
| Under $20,000 | 285 | 25.7 |
| $20,000–$39,999 | 277 | 25.0 |
| $40,000–$59,999 | 200 | 18.0 |
| $60,000–$79,999 | 156 | 14.1 |
| $80,000–$99,999 | 83 | 7.5 |
| $100,000 or more | 109 | 9.8 |

## Scale Items

The sample was asked a set of questions relating to their views toward privacy and the loyalty they possess toward companies and brands. Respondents were also asked questions relating to loyalty cards (their use of loyalty cards in everyday purchases, their understanding of the use of loyalty

**TABLE 1.2 Scale Items**

| |
|---|
| **Views toward Privacy** |
| Individual privacy is important to me. |
| Companies should never share personal information with other companies unless it has been authorized by the individuals who provided the information. |
| **Loyalty** |
| I am loyal to companies I trust. |
| I am loyal to the brands I trust. |
| **Opinions toward Loyalty Cards** |
| I use a loyalty card for most of my grocery purchases. |
| Loyalty cards are essentially "coupon cards" that minimize the need for coupons. |
| I read the user agreements of the loyalty cards I use. |
| **Trustworthiness of the Media** |
| The media is trustworthy. |

cards, and attention paid to the specifics of loyalty cards); their opinion of the trustworthiness of media; and several questions relating to their demographics. For questions not addressing demographics, respondents were asked to respond on a five-point scale, with 1 representing "disagree completely" and 5 representing "agree completely." The questions used are included in Table 1.2.

## RESULTS

First, the data were examined. A correlation table including each of the scale items is presented in Table 1.3. Each pair of items was observed to be significantly related (at the .05 level) except for the relationship between the item "Individual privacy is an important to me" and the item "The media is trustworthy."

As is displayed in Table 1.3, strong positive relationships (.391 to .685) are observed between the two privacy items and the two loyalty items. Support, therefore, is observed for the first hypothesis.

Given the strong relationships observed between the privacy and the loyalty items and the strong theoretical connection between concern for privacy and brand and store loyalty, the items were factor analyzed. The resulting one-factor solution accounted for 63.5% of the total variance. The factor loadings of the items ranged from .775 to .822. The Cronbach's Alpha for the four items is .807. The correlational results reported above, therefore, are further supported, providing further evidence in support of

**TABLE 1.3 Item Correlations and Levels of Significance**

| | Item 1 | Item 2 | Item 3 | Item 4 | Item 5 | Item 6 | Item 7 | Item 8 |
|---|---|---|---|---|---|---|---|---|
| **Item 1** | — | | | | | | | |
| **Item 2** | .685 | — | | | | | | |
| | .000 | | | | | | | |
| **Item 3** | .483 | .424 | — | | | | | |
| | .000 | .000 | | | | | | |
| **Item 4** | .438 | .391 | .653 | — | | | | |
| | .000 | .000 | .000 | | | | | |
| **Item 5** | .139 | .137 | .190 | .298 | — | | | |
| | .000 | .000 | .000 | .000 | | | | |
| **Item 6** | .180 | .158 | .273 | .379 | .567 | — | | |
| | .000 | .000 | .000 | .000 | .000 | | | |
| **Item 7** | .167 | .111 | .240 | .379 | .331 | .392 | — | |
| | .000 | .000 | .000 | .000 | .000 | .000 | | |
| **Item 8** | –.021 | –.071 | .231 | .161 | .162 | .207 | .293 | — |
| | .489 | .019 | .000 | .000 | .000 | .000 | .000 | |

Item 1 Individual privacy is important to me.
Item 2 Companies should never share personal information with other companies unless it has been authorized by the individuals who provided the information.
Item 3 I am loyal to companies I trust.
Item 4 I am loyal to the brands I trust.
Item 5 I use a loyalty card for most of my grocery purchases.
Item 6 Loyalty cards are essentially "coupon cards" that minimize the need for coupons.
Item 7 I read the user agreements of the loyalty cards I use.
Item 8 The media is trustworthy.

the first hypothesis. Individuals who have privacy concerns tend to be loyal to companies and brands.

When the correlations between the privacy items and the loyalty card items are examined, fairly consistent results were observed. The correlations between each of the factors were significant (at the .05 level) and positive. The correlations coefficients ranged between .111 and .379. Respondents who are more concerned with privacy were observed to be more likely to use loyalty cards for their grocery purchases and were more likely to view loyalty cards as "coupons cards" than were respondents who are less concerned with privacy, contrary to what was hypothesized. Respondents who are more concerned with privacy were observed to be more likely to read loyalty card user agreements than were respondents who are less concerned with privacy, consistent with what was hypothesized. The results, therefore, do not lend support to Hypothesis 2. Although individuals who are more concerned about privacy were observed to be more apt to read loyalty card user agreements, they expressed that they were more apt to use a loyalty card for grocery purchases and were more apt to view loyalty cards as coupon cards.

When the correlations between the privacy items and the trustworthiness of media item are examined, inconsistent results were observed—a significant (at the .05 level) negative relationship was observed between the second privacy item (whether companies should share personal information), consistent with Hypothesis 3. No significant relationship, however, was observed with the other privacy item (individual privacy is important to me). Limited support, therefore, was observed for Hypothesis 3.

Finally, when the relationships between individuals' views toward privacy and demographics are examined, inconsistent results are observed (see Table 1.4). When gender and age are examined, significant (at the .05 level) relationships were observed in a direction consistent with the Hypotheses 4a and 4b. Females were observed to be more likely to be concerned about privacy than males. Also, individuals for whom privacy is a concern are significantly older than individuals for whom privacy is a lesser concern.

When education and income were examined, however, very different results were observed. Significant (at the .05 level) relationships were not observed between individuals' views toward privacy and either education or income. Support, therefore, was not observed for Hypotheses 4c and 4d.

The analysis was extended to further understand the results. To do so, cluster analysis was conducted on the study participants based on the four privacy and loyalty items. The results consisted of two groups of individuals. The Silhouette measure of cohesion and separation exceeded .5, indicating good cluster quality. The sizes of the groups were similar, with the larger group (the second group) accounting for 54.7% of the sample, resulting in a ratio of sizes of 1.21, which is also regarded as good. The first group consisted of individuals

**TABLE 1.4 Concern for Privacy and Demographics: Item Correlations and Levels of Significance**

| | Individual Privacy Is Important to Me | Companies Should Never Share Information[a] |
|---|---|---|
| Gender | .117[b] | .137[b] |
| | .000 | .000 |
| Age | .170[c] | .227[c] |
| | .000 | .000 |
| Education | .000[c] | .029[c] |
| | .988 | .342 |
| Income | –.034[c] | .024[c] |
| | .257 | .426 |

[a] Entire Item: Companies should never share personal information with other companies unless it has been authorized by the individuals who provided the information.

[b] Cramer's V

[c] Spearman's Rho

for whom privacy is a greater concern and who are more loyal to companies and brands than the individuals who comprised the second group.

To provide insight into the differences between the two groups, t-tests, Mann-Whitney tests, and a chi-square test were conducted between the two groups on the other items— on individuals' opinions toward loyalty cards, their opinion of the trustworthiness of media, and several demographic factors (items examined in Hypotheses 2–4). The results of the tests are displayed in Table 1.5.

When the clusters formed based on the privacy and loyalty items are examined, significant (at the .05 level) differences were observed between the groups on each of the loyalty card items. The direction of the differences was observed to be consistent with those seen in the earlier analyses. Further support, therefore, was observed for the results noticed when Hypothesis 2 was tested.

When perceptions of the trustworthiness of the media across the two clusters is examined, a significant (at the .05 level) difference was observed between the groups. Interestingly, the difference is in the direction opposite of Hypothesis 3. The analysis of the clusters suggests that the process of combining views toward privacy and loyalty toward stores and brands affects the relationship with perceptions of the trustworthiness of the media.

**TABLE 1.5 Differences between the Two Groups of Respondents**

| | | Means | Std. Dev. | t-Value | Significance |
|---|---|---|---|---|---|
| Item 1 | Group 1 | 3.71 | 1.365 | 6.855 | .000 |
| | Group 2 | 3.20 | 1.111 | | |
| Item 2 | Group 1 | 3.69 | 1.182 | 8.147 | .000 |
| | Group 2 | 3.17 | 0.979 | | |
| Item 3 | Group 1 | 3.42 | 1.306 | 8.128 | .000 |
| | Group 2 | 2.85 | 1.002 | | |
| Item 4 | Group 1 | 2.68 | 1.239 | 3.336 | .001 |
| | Group 2 | 2.46 | 0.909 | | |
| Gender | Group 1 | NA | NA | Chi-Square = 11.537 | .000 |
| | Group 2 | NA | NA | | |
| Age | Group 1 | 3.72 | 1.642 | Mann-Whitney U Test | .000 |
| | Group 2 | 3.25 | 1.524 | | |
| Education | Group 1 | 3.43 | 1.267 | Mann-Whitney U Test | .824 |
| | Group 2 | 3.46 | 1.282 | | |
| Income | Group 1 | 2.87 | 1.619 | Mann-Whitney U Test | .254 |
| | Group 2 | 2.76 | 1.583 | | |

Item 1 I use a loyalty card for most of my grocery purchases.
Item 2 Loyalty cards are essentially "coupon cards" that minimize the need for coupons.
Item 3 I read the user agreements of the loyalty cards I use.
Item 4 The media is trustworthy

When gender and age are examined, significant (at the .05 level) differences were observed between the two clusters in a direction consistent with the previous analysis. Females were observed to be more likely to be in the cluster consisting of individuals for whom privacy is a greater concern. Also, the ages of individuals in the cluster consisting of individuals for whom privacy is a greater concern are significantly older than individuals in the other cluster. Again, consistent with the previous analysis, when education and income were examined, significant (at the .05 level) differences were not observed between the two clusters.

## DISCUSSION

With advances in information-gathering technology, businesses and organizations arguably have access to more information than ever before. In an arguably increasingly competitive business environment, additional information on one's customers can prove to be critical to ongoing business success. In response to increased efforts to gather information, however, concerns about a growing lack of privacy are increasing.

Due to their position at the end of the marketing channel—the only type of business that directly and consistently interacts with the final consumers—retailers have more access to information from consumers than any other member of the marketing channel. Moreover, the highly competitive nature of retailing also prompts retailers to be on the forefront of gathering and using information from consumers to improve their competitive positions. A primary source of information on their customers for many retailers is loyalty cards: Through the use of loyalty cards, retailers can record and retain data on all of the purchases made from their stores for each member of their loyalty programs. The use of loyalty card programs, however, has raised particular privacy concerns. The focus of this study is to add to the present knowledge base by examining the relationships between privacy and loyalty card programs for a large cross-section of the U.S. population.

The first hypothesis examined the relationship between views toward privacy and brand and store loyalty. The results suggest that these concepts are strongly related. The results suggest that individuals who are more concerned about privacy are more likely to exhibit both brand loyalty and store loyalty than individuals who are less concerned about privacy. Consumer packaged goods (CPG) companies and retailers, therefore, may find it easier to build loyalty among consumers who are more concerned about privacy. The findings also suggest that CPG companies and retailers may also find it advantageous to build concern about privacy among their present customers since by doing so they may be able to build additional loyalty. Clearly, additional research is required before such a strategy can be advocated, however.

The second hypothesis examined the relationship between views toward privacy and actions and opinions relating to loyalty cards. The results were mixed. Individuals who are more concerned about privacy were more likely to read user loyalty card user agreements, as would be expected. Surprisingly, however, they were also more likely to view loyalty cards as "coupon cards" than individuals who were less concerned about privacy. Is this due to how the privacy sections in the agreements are worded? The privacy sections are obviously written to be true, but they are also written in ways to assuage the privacy concerns of individuals. Does reading the agreements lower individuals' privacy fears, making the individuals who read the agreements more likely to view loyalty cards primarily merely as a means to receive discounts? Research on the nature of loyalty card agreements and their effect on individuals' privacy apprehensions about the card would seem to warrant research attention.

Even more surprising, individuals who express more concern for privacy were observed to be more likely to use loyalty cards for most of their grocery purchases. Hence, individuals with higher privacy concerns were observed to be more likely to share their personal grocery shopping behavior with their supermarkets. This counter-intuitional finding suggests the need for additional research. Does this finding result from the increased loyalty that individuals with higher privacy concerns tend to exhibit? Does their increased likelihood to be store loyal lead to increased trust toward the stores to which they are loyal? If individuals who are store loyal are more apt to use loyalty cards, the data which retailers obtain from these individuals will tend to be more useful and more valuable since it will provide retailers with a larger picture of their purchasing activities.

The third hypothesis examined the relationship between views toward privacy and perceptions of trustworthiness of the media. The results are inconclusive, although a relationship was observed for the use of data but not with the importance placed on personal privacy. Furthermore, when privacy and loyalty are viewed together, the results appear to be reversed. These findings suggest that concerns for privacy may not be part of an overall view of distrust toward institutions. Additional research into the basis and domain of individuals' views toward privacy appears to be warranted.

Finally, the fourth hypothesis examined the relationship between views toward privacy and demographics. Expected relationships were observed for gender and age. Females and older individuals were observed to be more concerned about privacy than were males and younger individuals. No relationships were observed with education and income, however. The lack of relationships with education and income suggests that concerns toward privacy may be independent of individuals' socioeconomic standing. Additional research into the demographic foundations of a concern for privacy appears valuable.

## Limitations

The inherent exploratory nature of the study reflects the study's primary shortcoming. Although the study is based on a sizable sample, its goal is to provide insight in the relationships between consumers' desire for privacy and their opinions about loyalty cards and their use. The study raises a number of potentially valuable areas for future research. Specifically, the findings from this study suggest that the relationship between individuals' views towards privacy and their use of loyalty cards may be much more complicated than may be commonly believed. Given the widespread use of loyalty cards, this study provides for several areas for future research.

## REFERENCES

Aitken, S. (2007). Are there really hidden dangers to data collection? *Precision Marketing, 19*(8), 12.

Allaway, A., Gooner, R. M., Berkowitz, D., & Davis, L. (2006). Deriving and exploring behavior segments within a retail loyalty card program. *European Journal of Marketing, 40*(11/12), 1317–1339.

Allen-Castellitto, A. (1999). Coercing privacy. *William & Mary Law Review, 40*(3), 723–757.

Batislam, E. P., Denizel, M., & Filiztekin, A. (2007). Empirical validation and comparison of models for customer base analysis. *International Journal of Research in Marketing, 24*(3) 201–209.

Bélanger, F., & Crossler, R. E. (2011). Privacy in the digital age: A review of information privacy research in information systems. *MIS Quarterly, 35*(4), 1017–1041.

Bellizzi, J. A., & Bristol, T. (2004). An assessment of supermarket loyalty cards in one major U.S. market. *Journal of Consumer Marketing, 21*(2), 144–154.

Berry, L. L. (1995). Relationship marketing of services: Growing interest, emerging perspectives. *Journal of Academy of Marketing Science, 23*(4), 236–245.

Burns, D. J., & Toncar, M. (2011). Involvement in retailer loyalty card programs: An exploratory look at consumers' attitudes and privacy concerns. In C. David Strupeck (Ed.), *Academy of Business Disciplines Proceedings* (Section 22). Floyds Knobs, IN: Academy of Business Discipline.

Capizzi, M. T., & Ferguson, R. (2005). Loyalty trends for the twenty-first century. *Journal of Consumer Marketing, 22*(2), 72–80.

Chaudhuri, A., & Holbrook, M. B. (2001). The chain of effects from brand trust and brand affect to brand performance: The role of brand loyalty. *Journal of Marketing, 65*(2), 81–93.

Cortinas, M., Elorz, M., & Mugica, J. M. (2008). The use of loyalty-card databases: Differences in regular price and discount sensitivity in the brand choice decision between card and non-card holders. *Journal of Retailing and Consumer Services, 15*(1), 52–62.

Davenport, T. H., & Harris, J. G. (2007). The dark side of customer analytics. *Harvard Business Review, 85*, 37–48.

Demoulin, N. T. M., & Zidda, P. P. (2008). On the impact of loyalty cards on store loyalty: Does the customers' satisfaction with the reward scheme matter? *Journal of Retailing and Consumer Services, 15*(5), 386–398.

Dick, A. S., & Basu, K. (1994). Consumer loyalty: Toward an integrated conceptual framework. *Journal of the Academy of Marketing Science, 22*(2), 99–113.

Du, R. Y., Kamakura, W. A., & Mela, C. F. (2007). Size and share of customer wallet. *Journal of Marketing, 71*, 94–113.

Duhigg, C. (2012, February 12). How companies learn your secrets. *New York Times Magazine.* Retrieved from http://www.nytimes.com/2012/02/19/magazine/shopping-habits.html?pagewanted=1&_r=2&hp

Felgate, M., Fearne, A., Di Falco, S., & Garcia Martinez, M. (2012). Using supermarket loyalty card data to analyse the impact of promotions. *International Journal of Market Research, 54*(2), 221–240.

Gavison, R. (1980). Privacy and the limits of law. *Yale Law Journal, 89*(3), 421–471.

Geçti, F., & Zengin, H. (2013). the relationship between brand trust, brand affect, attitudinal loyalty and behavioral loyalty: A field study towards sports shoe consumers in Turkey. *International Journal of Marketing Studies, 5*(2), 111–119.

Halim, R. E. (2006). *The effect of the relationship of brand trust and brand affect on brand performance: An analysis from brand loyalty perspective (A case of coffee instant product in Indonesia).* Working paper, Universitas Indonesia, Graduate School of Management. Retrieved from http://ssrn.com/abstract=925169

Ibáñez, V. A., Hartmann, P., & Calvo, P. Z. (2006). Antecedents of customer loyalty in residential energy markets: Service quality, satisfaction, trust and switching costs. *The Service Industries Journal, 26*(6), 633–650.

Klabjan, D., & Pei, J. (2011). In-store one-to-one marketing. *Journal of Retailing and Consumer Services, 18*(1), 64–73.

Laczniak, G. R., & Murphy, P. E. (2006). Marketing, consumers and technology: Perspectives for enhancing ethical transactions. *Business Ethics Quarterly, 16*(3), 313–321.

Lohr, S. (2010, January 3). A data explosion remakes retailing. *New York Times.* Retrieved from http://www.nytimes.com/2010/01/03/business/03unboxed.html?_r=0

Liu, C.-T., Guo, Y. M., & Lee, C.-H. (2011). The effects of relationship quality and switching barriers on customer loyalty. *International Journal of Information Management, 31*(1), 71–79.

Madden, M., Fox, S., Smith, A., & Vitak, J. (2007). *Digital footprints: Online identity management and search in the age of transparency.* PEW Research Center Publications. Retrieved from http://pewresearch.org/pubs/663/digital-footprints

Magi, A. W. (2003). Share of wallet in retailing: The effects of customer satisfaction, loyalty cards and shopper characteristics. *Journal of Retailing, 79*(2), 97–106.

Mauri, C. (2003). Card loyalty: A new emerging issue in grocery retailing. *Journal of Retailing and Consumer Services, 10*, 13–25.

Milne, G. R., & Bahl, S. (2010). Are there differences between consumers' and marketers' privacy expectations? *Journal of Public Policy & Marketing, 29*(1), 138–149.

Mohammad, A. A. S. (2012). The Effect of Brand Trust and Perceived Value in Building Brand loyalty. *International Research Journal of Finance & Economics, 85*, 111–126.

Nissenbaum, H. (2010). *Privacy in context: Technology, policy, and the integrity of social life.* Stanford CA: Stanford University Press.

Omar, N. A., Alam, S. S., Aziz, N. A., &Nazri, M. A. (2011). Retail loyalty programs in Malaysia: The relationship of equity, value, satisfaction, trust, and loyalty among cardholders. *Journal of Business Economics & Management, 12*(2), 332–352.

O'Neil, D. (2001). Analysis of internet users' level of online privacy concerns. *Social Science Computer Review, 19*(1), 17–31.

Palmer, A., McMahon-Beattie, U., & Beggs, R. (2000). Influences on loyalty programme effectiveness: A conceptual framework and case study investigation. *Journal of Strategic Marketing, 8*(1), 47–66.

Passingham, J. (1998). Grocery retailing and the loyalty card. *Journal of the Market Research Society, 40*(1), 55–63.

Pauler, G., & Dick, A. (2006). Maximizing profit of a food retailing chain by targeting and promoting valuable customers using loyalty card and scanner data. *European Journal of Operational Research, 174*(2), 1260–1280.

Payne, D., & Trumbach, C. (2009). Data mining: Proprietary rights, people and proposals. *Business Ethics: A European Review, 18*(3), 241–252.

Peppers, D., & Rogers, M. (2004). *Managing customer relationships: A strategic framework.* Hoboken, NJ: John Wiley.

Robson, J. S. P. (2006). Embrace the discount card. *Fraser forum special issue on labour regulation and unionization,* (May), 32.

Schermer, B. W. (2011). The limits of privacy in automated profiling and data mining. *Computer Law & Security Review, 27*(1), 45–52.

Sharp, B., & Sharp, A. (1997). Loyalty programs and their impact on repeat-purchase loyalty programs. *International Journal of Research in Marketing, 14*(5), 473–486.

Smith, A., Sparks, L., Hart, S., & Tzokas, N. (2003). Retail loyalty schemes: Results from a consumer diary study. *Journal of Retailing and Consumer Services, 10,* 109–119.

Smith, A. D. (2008). Customer loyalty card programmes and the interaction with support technology in the retail industry. *International Journal of Management and Enterprise Development, 5*(2), 157–195.

Smith, R. E. (2000). *Ben Franklin's web site: Privacy curiosity from Plymouth Rock to the Internet.* Providence RI: Privacy Journal.

Solove, D. (2006). A taxonomy of privacy. *University of Pennsylvania Law Review, 154*(3), 477–564.

Sung, Y., & Kim, J. (2010). Effects of brand personality on brand trust and brand affect. *Psychology & Marketing, 27*(7), 639–661.

"Survey: Satisfying and Retaining Customers Is Top Priority." (2009). *Chain Store Age, 85*(3), 18.

Van den Hoven, J. (2007). Information technology, privacy and the protection of personal data. In J. Van den Hoven (Ed.), *Information technology, privacy and the protection of personal data* (pp. 462–494). New York, NY: Cambridge University Press.

Wacks, R. (2010). *Privacy: A very short introduction.* New York, NY: Oxford University Press.

Warren, S., & Brandeis, L. (1890). The right to privacy (the implicit made explicit). In F. D. Schoeman (Ed.), *Philosophical dimensions of privacy: An anthology* (pp. 193–220). Cambridge MA: Harvard Law Review.

Westin, A. (1967). *Privacy and freedom.* New York, NY: Atheneum.

Zeher, C., Şahin, A., Kitapçi, H., & Özşahin, M. (2011). The effects of brand communication and service quality in building brand loyalty through brand trust: The empirical research on global brands. *Procedia: Social and Behavioral Sciences, 24,* 1218–1231.

Ziliani, C., & Bellini, S. (2004). Retail micro-marketing strategies and competition. *International Review of Retail, Distribution and Consumer Research, 14*(1), 7–18.

CHAPTER 2

# IDENTIFYING PROFITABLE CUSTOMERS USING A TWO-STAGE LOGISTIC MODEL

## An Application from B2B Credit Card Marketing

**Vernon Gerety**
***Westchester State University***

**Stephan Kudyba**
***New Jersey Institute of Technology***

A successful marketing campaign should be measured not simply by generating new customers but by its ability to generate profitable customers. However, a marketer's challenge is multifaceted in the credit card industry. It involves not only identifying the prospect market and getting an offer to the prospect that is designed to entice the customer to apply for credit, but also, when extending credit, getting the applicant approved and activated. Ultimately, success is driven by not only increasing active accounts but also increasing the number of customers who utilize the service that is profit-

*Contemporary Perspectives in Data Mining, Volume 2*, pages 25–32

able to the credit card issuer.[1] The combination of data and analytic methods can help marketers achieve these goals.

In this chapter, we address the business process of marketing prepaid credit cards to small businesses. Data corresponding to potential customers are analyzed, which incorporates a two-stage logistic model to isolate and identify prospects and customer profitability. This two-stage process is essentially a conditional probability exercise, where we estimate the probability of high profits given card activation. This approach is particularly effective in this case that involves analysis of small companies, since data availability is limited. In fact, given the data limitations, it will be demonstrated that this two-stage approach for identifying profitable customers is superior to simply building a single model to predict a prospect's profitability or a prospect's likelihood to respond.

## PROBABILITY THEORY

The Kolmogorov definition of conditional probability is a fundamental concept in statistics, typically denoted as

$$P(A|B) = \frac{P(A \cap B)}{P(B)}$$

where $P(A|B)$ denotes the conditional probability of A given B. Assuming events A and B are not independent of each other, this conditional probability equals the joint probability of A and B divided by the probability of B.

Bayesian inference is a dynamic way to interpret conditional probabilities over time and has been demonstrated to be effective in limited marketing applications (Wei, 2007). Although theoretically appealing, it is difficult to design and implement, and it has limitations in applications to practical business problems.

A more developed and widely used technique is logistic regression. Applications of logistic regression in predicting categorical data have advanced significantly in the last 15 years (Stokes, 1999). However, when it comes to using this technique, practical applications using conditional probability seem limited.

This chapter proposes a simple and effective way to estimate conditional probabilities that is easy to implement and deliver effective solutions for many broad business applications—in this case, direct marketing.

## BUSINESS APPLICATION

The data for this analysis were provided by a financial services company offering prepaid credit cards to small to medium-sized businesses. For this market, small to medium-sized business were defined as firms with 6 to 50 employees or annual sales between $1 and $10 million.

The institution had identified a target market by applying market research techniques. The prepaid business card offered an effective means for targeted business customers to manage their expenses of employees who need funds during their normal job responsibilities (e.g., sales persons, deliver drivers, executive secretaries, etc.). The company would simply distribute cards to corresponding employees with a set dollar amount on it. The prepaid card in this case provides the targeted business with a much larger amount of funds available (virtually any amount can be deposited on the prepaid cards) compared to what would be available if the same firm attempted to get an approved line of credit on a business credit card.

The development of this model was challenged by the fact that the target audience for the clients' product was small to mid-size businesses. Business to business (B2B) marketers need to deal with both limited and incomplete data for prospects of this size. In general, the B2B analytical challenge in this sector is having limited or missing data, where the premium is on using creative analytical techniques to enhance predictive performance and successful applications.

## IDENTIFYING PROSPECTS WITH HIGH PROFIT POTENTIAL AS CUSTOMERS

As suggested above, this chapter illustrates a practical methodology to estimate conditional probability. The goal is to identify prospects that are more likely to become profitable customers. The analytical method described will be denoted as a two-stage logistic model. The objective function is the conditional probabilities of a profitable customer given funding. The analytic process was developed to increase the marketing effectiveness of an organization providing prepaid credit cards to a target market—more specifically, to identify prospects that will have a high likelihood of not only applying for the card but also to fund *and* eventually become a profitable customer.

## OBJECTIVE FUNCTION

In this section we provide more details on the dataset used for the analysis and the "items of interest" or business behavior we are attempting to predict.

As mentioned above, the product and issuer were new to the market; therefore, the prospecting efforts at time of the analysis were a bit haphazard. The good news from a statistical perspective is the dataset did provide a nice random sample of the prospect universe, which provides more variance to the data set and hence more opportunity to identify key metrics of success.

Three data groupings were created:

1. *Prospect Universe*—based upon client's website traffic and traditional marketing efforts along with Dun &Bradstreet marketing database (see discussion below)
2. *Fund Customers*—customers who applied for card, were approved, and funded their cards
3. *Top Customers*—customers who generate the highest level of profits based upon a ranking of all customers using activity from the previous 12 months of customer's account usage

Dun & Bradstreet (D&B) marketing attributes were matched to all prospects and customers. These data along with customer behavior data were used to create the dependent and independent variables, and this dataset was used to conduct the bivariate analysis and statistical modeling.

Data elements analyzed included a business credit risk score, number of employees, business age (time in business), industry segment (SIC code), number of business trades reported, total sales, number of report inquiries for the business,[2] and number of website addresses (defined as a count of unique URLs) identified for the business.

Using this dataset, two dependent variables were created:

1. FUND_IND: A dummy variable, companies who were approved and funded flagged as FUND_IND = 1, else FUND_IND = 0.
2. TOPCUST_IND: A dummy variable, for customers whose profitability exceeds a hurdle defined by the financial institution as TopCust_IND = 1, else TopCust_IND = 0.

## METHODOLOGY

Logistic regression using SAS was utilized to estimate the statistical models. Using the two dependent variable, a series of 3 algorithms were estimated where D&B attributes and customer behavior data were used as independent variables.

Three models/algorithms were estimated and compared.

1. Funded Customer Model: Using the *FUND_IND* variable as the dependent variable, this estimated the probability of prospect activating or funding their cards, *P(Fund)*.

2. *TopCust Model*: Using *TOPCUST_IND* as dependent variable and the p value from the Funded Model as one of the dependent variables along with other attributes, this second stage model estimated the conditional probability p value *P(TopCust|Fund)*.

## RESULTS

The results are presented below and reveal the following:

1. Overall statistical power of both models is quite high, as measured by predictive lift by ranked ordered decile. See Figure 2.1.
2. The lift from the TopCust model is significant relative to the model predicting Funded only. See Figure 2.2.
3. Benefits from focused prospecting are measured in terms of fewer qualified prospects and higher yields from marketing efforts. See Table 2.1.

Figure 2.1 can be interpreted as follows: Each decile represents 10% of the sample ranked ordered from highest probability to lowest based upon the respective predictive algorithm, P(TopCust) versus P(TopCust|Fund); for each decile the percentage of TopCust were calculated. For example, in decile 1 the ranking based upon *P(TopCust|Fund)* contained 53% profitable customers (TopCust=1), while for the *P(Fund)* the percentage of profitable customers was 32.8%.

Alternatively, Figure 2.2 focused on the cumulative number of profitable customers identified by each model. For example, in the top 2 deciles, the

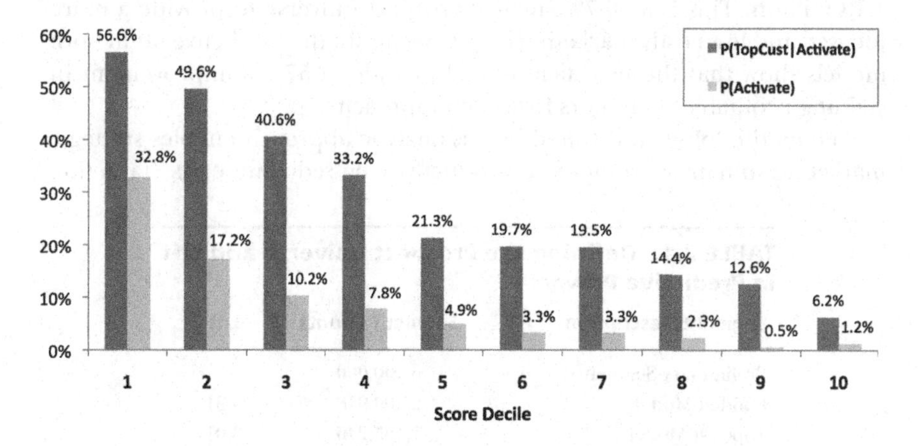


**Figure 2.1** Percentage of top customers identified by score decile comparing funded versus top cust conditional models.

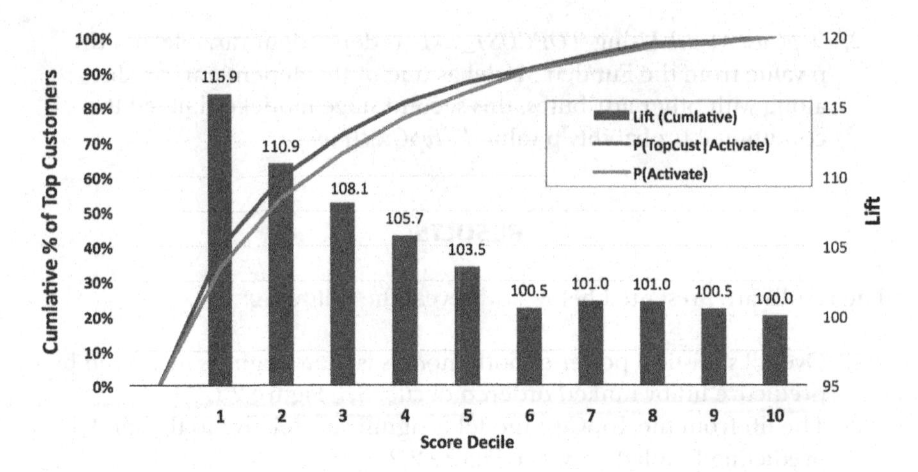


**Figure 2.2** Cumlative top customers by declie comparing funded vs top cust conditional models.

ranking based upon *P(TopCust|Fund)* contained a total percentage of profitable customers (TopCust=1) equals 60%, while for the *P(Fund)* the total percentage of profitable customers (TopCust=1) equals 54%, which is 11% higher total, or a lift of 110.9.

By reviewing Table 2.1, it is demonstrated that segmentation using simple selectors such as SIC and sales volume (*Preliminary Segments*) enabled the client to identify a market of 6 million prospects. Using the Funded Model, the client was able to narrow this segment to a little bit over 2.4 million prospects, and finally using the two-stage model to identify "top customers" based upon profitability, the prospect universe narrowed to 1.18 million. This is a 50.7% smaller prospect universe to provide a more focused and less costly marketing effort. Secondly, the predictive lift in both models show that the two stage model provides a 57.7% improvement in yielding profitable customers from this approach.

The methodology illustrated in this analytic approach enables strategic marketing initiatives to increase productivity by reducing costs (targeting

**TABLE 2.1 Defining the Prospect Universe and Lift in Predictive Power**

| Segment Description | Segment Counts | Lift |
|---|---|---|
| Preliminary Segments | 6,000,000 | |
| Funded Model | 2,404,918 | 1.91 |
| TopCust Model | 1,185,246 | 3.01 |
| Improvement | –50.7% | 57.7% |

a smaller buy more focuses segment) while yielding a higher percentage of profitable customers. Here "lift" is defined as the increase in the rate of top customers identified relative to the average rate from the entire prospect universe of 6MM names. This lift is calculated for both the *Funded* model and the two-stage *TopCust* model. The improvement is measured by improvement in lift for the TopCust model versus the Funded model.

## CONCLUSION

This chapter provides a sound platform to enhance efficiencies in marketing initiatives, which is particularly advantageous given the evolving state of the marketing spectrum as the information economy continues to progress. More traditional tactics of yesteryear, such as telemarketing and direct mail, are giving way to e-based tactics; however, as different as tactics such as text messaging to smartphones, email messaging, and social media interaction may be from the brick-and-mortar style, the core elements to marketing remain the same. The information age has created a data avalanche to both the businesses and the customer. Business is battling to get out its message to generate product awareness and eventually to convert awareness into paying customers. So some of the upside factors of the internet (increased reach, diverse contact, and interaction mechanisms) has the downside of increasing the battle for attention to your message. The result is that analytics play a vital role in managing these activities more than ever.

Above we discussed a two-stage approach to change the traditional marketing focus from new accounts to profitable customer. It was illustrated that this approach can provide significant lift in predictive power.

## NOTES

1. For a similar discussion from the hospitality industry see Morrison, Bose, and O'Leary (2000).
2. Business inquiry is based upon companies pulling a D&B report on the business. Typically this is consistent with a firm with a higher demand for credit. Not surprisingly, the analysis indicated a positive correlation with the likelihood of activating—that is, P(Fund) increases with number of business inquires.

## REFERENCES

Morrison, A. M., Bose, G., & O'Leary, J. T. (2000). Can statistical modeling help with data mining? A database marketing application for U.S. hotels. *Journal of Hospitality & Leisure Marketing, 6*(4), 91–110.

Stokes, M. E. (1999). *Recent advances in categorical data analysis.* Cary, NC: SAS Institute.

Wei, C. (2007). *Predicting customer responses to direct marketing: A Bayesian approach.* Unpublished doctoral dissertation, Lingnan University, Hong Kong.

CHAPTER 3

# A FRACTIONAL FACTORIAL ANALYSIS FOR IN-STORE PROMOTIONS

**Peter Charette**
***Waypoint Consulting, West Chester, PA***

**John Stanton**
***Saint Joseph's University***

**Neal Hooker**
***The Ohio State University***

## ABSTRACT

This chapter aims to provide an understanding of customers' attractiveness to promotions by using a fractional factorial analysis to develop an optimum display of various in-store promotional materials for mushrooms. The in-store promotional materials are combined into various collages that are developed with an incomplete bloc design and tested through video-simulated shopping approach. This analysis determined that certain promotional material would increase mushroom consumption over others. Additionally, certain promotional material had a varying effect on different demographics.

*Contemporary Perspectives in Data Mining, Volume 2*, pages 33–54

## INTRODUCTION

Many environmental factors cause shifts in shopper behavior throughout all industries. A change in behavior is usually accompanied with innovation and/or adjustments of marketing strategies. Current day innovations and adjustments are used to offer the shopper a better value. The food industry may be the most affected by environmental factors. As a result, the food retail industry has become a very competitive environment. It has become competitive for the manufacturer, the farmer, and the retailer. The rise in oil costs, droughts, and substitutes all put additional pressure on farmers to provide an affordable product. Consequently, due to the plethora of items supermarkets offer, it is not very difficult to find a substitute for any product. External pressures on the industry have given rise to operational costs that affect all the stakeholders involved. Various methods and strategies have been developed or tested in order to create a competitive advantage. In particular, advertising and promotions are used to increase total sales of a particular category.

Recent advances in the retail industry and the field of data analytics have provided extensive ability to measure success rates of each promotional campaign, thus creating an advantageous position for the firms that have analytic capabilities. Firms will continue to use the campaigns that work and stop wasting money on ineffective campaigns. However, when introducing a new product, it is impossible to proactively use data analytics to measure campaigns. Therefore, a campaign's success or failure will have to be put in the field in order to be measured.

The mushroom industry has encountered a variety of the aforementioned external hurdles, combined with the low popularity of the mushroom category. The mushroom industry thought offering a new product would help change consumer behavior. For most companies, new products are a popular growth strategy. Unfortunately, new products have proven to be time-consuming and very costly. In 2010, a total of 21,528 new food products were introduced in retail supermarkets (USDA, 2013). However, only 11% of these products had shown success (Linton & Demand Media, n.d.).

The mushroom industry has developed a new product that offers a solution to a problem that 70% of Americans have. The industry could use out-of-store advertising to make consumers aware of the new product, but the industry is comparably less financially endowed than other crops and CPG companies that already advertise out-of-store. With a significantly lower marketing budget, the industry needs to put its marketing dollars where they can have the greatest effect on consumers. Due to consumers' lack of attention towards mushrooms out of store, the best way to attract attention and increase purchases of mushrooms is at the point of purchase phase for shoppers. Thus, the purpose of this research is not only to increase the purchase volume of the mushroom category (both the new and existing product) but also to inform the shoppers of the benefits the new product can offer.

## LITERATURE REVIEW

The current state of growth for the mushroom industry has become stagnant. With overall prices continuing to increase, this indicates that total per capita consumption is falling. Mushrooms are a very elastic product. According to AC-Nielsen, a 10% price increase will decrease demand by 11.3%. A potential way to curtail this trend is to add a new value-added product that addresses consumers' needs. The mushroom industry identified a consumer problem: the abundant amount of Americans who have vitamin D deficiency. Over 75% of teens and adults have deficiency in vitamin D (Lite, 2009). The mushroom industry offers a solution with a new product. The mushroom industry's new value-added product is a mushroom with a 100% of a consumers' recommended daily amount (RDA) of vitamin D. The industry does not need to pass claims through the FDA and USDA because the process of increasing vitamin D is a naturally occurring component of mushrooms. The mushroom industry's new product could promote sales of the industry by using vitamin D promotional strategies that pull more consumers to the section. While vitamin D sales are expected to rise, the entire mushroom category is expected to rise along with it.

Point of purchase marketing (better known as shopper marketing) is a young practice. Many academics point out the need for future research in this field. Ailawadi, Beauchamp, Donthu, Gauri, and Shankar, (2009) discusses the need for research in store promotions, stating the need for a deeper understanding about the effectiveness of in-store advertising and promotions on shoppers. Additionally, how should a retailer allocate resources toward store coupons, feature advertising, and other in-store efforts for products? Kiran, Majumdar, and Kishore (2012) express further need of research in point of purchase phases:

> Manufacturers need information on the effectiveness of in-store stimuli and the extent to which they influence consumer purchasing behavior for their brands. Retailers also need this information to determine the effectiveness of these resources designed to stimulate additional sales and perhaps differentiating themselves from the competitors. (p. 37)

Both Steel (2008) and Kiran et al. (2012) show the potential contributions further research may provide. "In the latest effort of retailers, many of the in-store target advertising efforts are still in the early stages of development" (Steel, 2008, para. 6). This is a supportive statement for the proposed research on the presumed advancement in terms of in-store advertisement, which has to be explored. (Kiran et al., 2012).

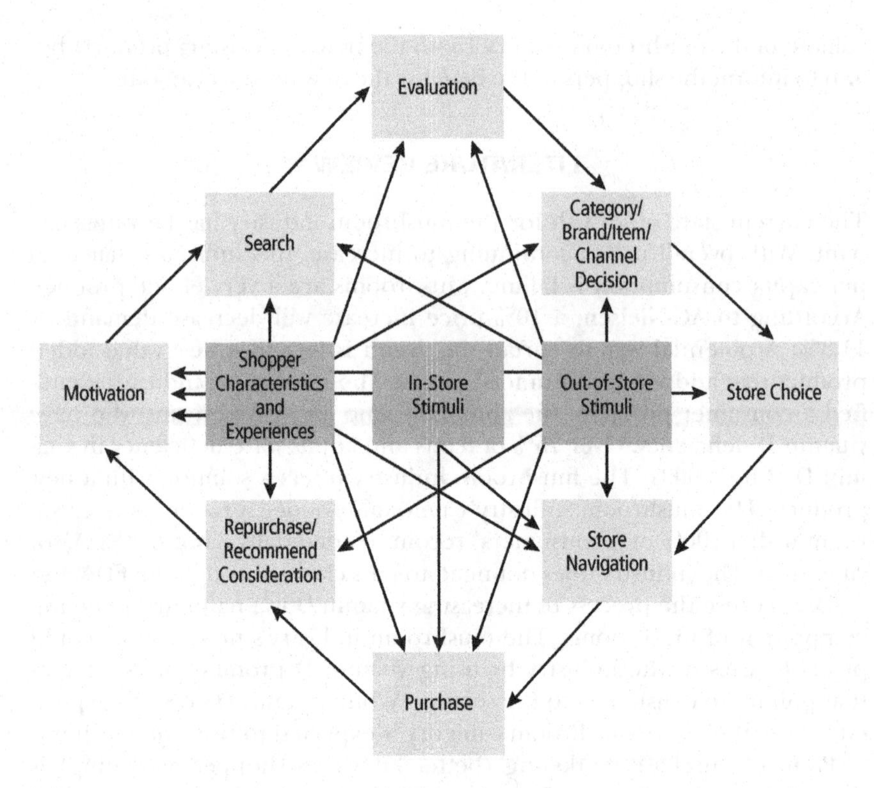


**Figure 3.1** Shopper behavior lifecycles.

A key goal of shopper marketing is to increase the likelihood of purchase. Shopper marketing is an acknowledgement of the need to understand, activate, and engage with consumers when they are in the role of shopper. A key focus of shopper marketing is to influence shoppers throughout the shopping cycle that comprises different stages such as motivation to shop, search, evaluation, category/brand/item selection, store choice, store navigation, purchase, repurchase, and recommendation considerations (Figure 3.1).

## METHODOLOGY

This research develops 32 different combinations of in-store promotions. The combinations can be varied by the content shown on each promotion, number of promotions present in a section, as well as type of promotional material. The varying content contains different levels of creativity in the design and size of signage, different marketing health claims, four varying slogans, and varying marketing pictures. The combination of promotional material will vary from 1 to 7 pieces throughout the various profiles. There are seven attributes; in this research the attributes are different types of promotional material. The seven different attributes each have three or more levels. Levels mean that there are different characteristics placed on each type of promotion. If there are three levels, then that means there are three interchanging characteristics placed on the attribute. No two levels of the same attribute will be shown at the same time. Only combinations of one attribute set of characteristics (levels) will be shown at a time. Each comparison contains varying amounts of attributes. The first attribute (promotional material) is in-store signage for the section, which has three different levels: simple, medium, and complex.

The second attribute is a shopping cart advertisement. A shopping cart advertisement is placed in the view of the section and is located on the front end of a cart—the end that faces forward as a consumer is pushing the cart. This attribute contains three levels: a food comparison that compares the research product mushrooms to milk and family pictures to explore marketing effects on attractiveness; additionally there was an empty slot with no advertisement to test whether or not it would be better to not have a shopping cart advertisement. The third attribute is a coupon. There are three levels of coupons: two with varying pictures and slogans and an empty slot. Additionally, the amount discounted is the same throughout the study. The fourth attribute is a shelf pop-out. A shelf pop is a promotional piece that is most easily seen from the side of the section. When a customer is walking down the aisle it should be perpendicular to their line of sight, giving them a clearer view. There are three levels of this attribute—two levels with varying claims, colors, and sizes—and the last level was an empty slot with no advertisement to test the effect of having the attribute.

The fifth attribute is a floor graphic. Floor graphics are images that are placed on the floor, typically in front of the section to pull in consumers to look at the section. This attribute has three levels: two levels with varying content and an empty slot with no advertisement to test the effect of having no advertisement. The sixth attribute is a recipe holder; it is a container that holds recipes that will incentivize customers to purchase the product based on the meal that could be prepared. This attribute has three levels: two with different content and an empty slot with no advertisement to test the effect of having no advertisement. The last and final attribute is the name of the new product; this attribute is placed within various other attributes. This attribute has four levels: two names that tested very highly in naming simulation, a name that tested very poorly in the same simulation, and no name to test the effectiveness of even having a name or not.

The research question will be answered by determining which combination of levels within the varying attributes creates the highest level of attraction for consumers. Note that the varying levels with no attribute in the slots also create the ability to see the optimum number of advertisements per section based on attractiveness.

## Interpreting a Conjoint Analysis

In a conjoint analysis, consumers' preference can either be rating based or ranking based. The preference is the dependent variable, and independent variables are the product attribute levels. The dependent variable can also be a binary preference intention (yes or no, a one-to-one comparison) or the dependent variable could be constant sum. The coefficients in the regression model are the estimated part-worth utilities. In analyzing a regression output, the R-square gives an indication on how well the data fit the model. R-square tells you the proportion of the variance of respondents' preferences explained by the combination of independent variables (attributes and levels). A conjoint analysis R-squares are inherently low; a low R-square would not have the equivalent negative connation as it would with other statistic methods.

Certain assumptions exist when evaluating a conjoint model. In the most basic terms, the profiles involved must make sense. The parameters must be based upon conceptual terms as well as practical issues. The independent variable values show how important an attribute is in affecting consumers' attraction to a promotion. Typically, if the coefficients were derived from parameters having varying ranges, the value would be derived from part-worth utilities for each attribute. If there were varying parameter ranges, to determine the utility of the part worth, one would subtract the smallest part-worth utility from the largest one. Then the total range is computed by adding all the ranges of the attributes together. The importance of each attribute is computed by dividing the range of the attribute by the total range, although in this case, all parameters are the same range (dummy coded); therefore, the transformation is not needed.

When validating a model, some practices suggest training a model with a sample of the respondents. Applying the regression coefficients derived from the trained model to predict value and actual values for a validation group. Additionally, one could score the results by adding a new population into the validation and training model to get the predicted value of the respondent's preferences of the dependent variable. If there is a high correlation between the training model and the validating models values with the score data, it will indicate that models have good predictive value. The scored model is used with separate observations typically referred to as holdout observations. In a conjoint analysis, the observations for developing a validating model are used with actual observations. Holdout observations are used to validate the model and to calibrate the simulator. The profile for validation is a hypothetical product that is rated or ranked by consumers but is not used in the estimation of utility values in the model. The purpose of having holdout observations is to determine internal validity of the model by examining the associations between the actual and predicted ratings for these observations. Usually, the number of holdouts is small because these profiles, though not used in estimating the model, add to the burden placed on respondents. It is not necessary to increase the burden on consumers when they are performing tasks.

These holdout observations are normally derived from those that are included in a full-factorial design but not in fractional factorial design. In this research, instead of using a traditional holdout method as validation, it was replaced by doing both a choice base and rating base, and cross-referenced the analyses. Another indicator for the assessment of the model would be the p-value associated with the level of the significance (e.g., .05 or .01). However, for a conjoint model, the more profiles a design contains, the more significant each profile's p-value needs to be according to the protected error rate. For example, in this research there are 32 profiles, in order to achieve significant value for the conjoint model at a p-value of 0.1; each attribute's significance must be equal to or less than .00328 {1–(1–32)^(1/0.1). To calculate the point of sale effectiveness, a survey was conducted, comprised of trial and usage data, and demographics were collected and analyzed using traditional methods and/or replicating the work of other produce promotions in academia and business. The survey largely consisted of a conjoint or trade-off analytic experiment where respondents were asked to rate their attraction to various levels of in-store promotional material. The conjoint experiment quantified the relative importance of and tradeoffs between the type of promotions and the differing characteristics of content, marketing appeal, and creativity on each promotion. Price was held constant throughout the survey; each profile of promotions was tested against a standard (control). The standard consisted of a mushroom section in the highest-rated chain supermarket in the greater Philadelphia area: ShopRite.

A fractional factorial design was used in order to develop the profiles for the study. The fractional factorial analysis is due to the large array of materials that are being tested. For example, if there was no fractional factorial design utilized (full factor analysis), there would be: (3 × 3 × 3 × 3 × 3 × 3 × 4) 2,916 possibilities of combinations. In order to test that many combinations, you would need a considerably large sample size and an extremely large amount of money to have any statistical significance. Therefore, a fractional factorial design is needed in order to make the analysis of these factors feasible. Using SPSS (conjoint add-on), a fractional factorial analysis is developed. A fractional factorial analysis is able to reduce the design down to 32 profiles (combinations of material needing to be tested), compared to the 2,916 in a full factorial design. A fractional factorial analysis eliminates the possibility of independent variables being multi-collinear. Multi-collinear variables in many cases increases values (beta coefficients) unrealistically as well as inflates the R-square value (error rate), thus making the model inaccurate.

The fractional factorial design is developed using the two levels of promotional material, along with testing the absence of a claim. The absence is tested to understand whether or not the section is more attractive with no claim of this type or none at all. All attributes will have an absence level except signage. Signage does not have an absent level because logically that would not make sense. Consumers need to know what the product is, and the price point must be displayed. Additionally, no properly run supermarket would knowingly have an absence of any sign.

The terminology for describing a conjoint design is as follows: An attribute is a section of like material that is being tested—a category. In this research the attributes are the different types of promotions (signage, coupons, shopping cart advertisements, floor graphics, name of product, and shelf popouts). Levels are under each attribute but have varying characteristics. For example, the attribute of signage has three levels: complex signage, moderate signage, and simple signage. All of these different levels of signage have varying characteristics that are going to be weighted to conclude which is the most attractive to consumers. Profile or card consists of the combinations of the different levels of attributes being tested; in this case there are 32 different profiles (i.e., 32 combinations of promotional material). Only one attribute's level will be tested per profile (see Table 3.1).

The selection of the standard was a decision to do the cross-validity check was an attempt to offer a new way of conducting the conjoint analysis. Each profile is tested against the same criteria, the standard. The standard offers another type of validity known as ecological validity. Ecological validity as defined by McKechnie (1977) refers to the applicability of the results of laboratory analogues to non-laboratory, real life settings. The development and validation of environmental simulations are important for the rapprochement of internal and external validity. Bosselmann and Craik (1987) have reported significant association between direct and simulation presentations as a result of ecological validity.

Shankar, who was mentioned earlier, called for further research to better understand shoppers' behavior. The development of each profile was designed to contribute to this field by developing a more interactive approach trying to simulate the actual supermarket experience. Each card's output taken from SPSS (conjoint add-on) output was Photoshopped into an actual supermarket's mushroom section. The actual supermarket section was the ShopRite in Seville, New Jersey. There were four pictures taken of the section: one from afar as a shopper enters the produce section, one from the side as if a shopper was walking perpendicular to the section, another from the front which shows the entirety of the mushroom section, and one of the floor in front of the section. Each of the four pictures was

**TABLE 3.1**

| Card ID | Floor Graphic | Recipe Holder | Shelf Pop-out | Coupon | Signage Levels | Shopping Cart | Vitamin D Name |
|---|---|---|---|---|---|---|---|
| 1 | None | Recipe 2 | Pop 2 | Coupon 2 | Simple | Shopping Cart Ad 2 | No Name |
| 2 | ad 2 | Recipe 1 | Pop 2 | None | Simple | Shopping Cart Ad 1 | BAD NAME |
| 3 | ad 1 | None | Pop 2 | Coupon 1 | Moderate | None | No Name |
| 4 | ad 1 | None | Pop 1 | Coupon 1 | Simple | Shopping Cart Ad 1 | Sun Enriched |
| 5 | ad 1 | Recipe 2 | None | Coupon 2 | Complex | None | BAD NAME |
| 6 | ad 2 | None | None | None | Simple | Shopping Cart Ad 1 | No Name |
| 7 | ad 2 | None | None | None | Complex | None | Sun Enriched |
| 8 | None | None | Pop 2 | None | Moderate | None | BAD NAME |
| 9 | ad 1 | Recipe 2 | None | None | Simple | None | Sun Enriched |
| 10 | None | Recipe 2 | None | coupon 1 | Moderate | Shopping Cart Ad 1 | BAD NAME |
| 11 | None | None | Pop 1 | coupon 2 | Complex | Shopping Cart Ad 1 | No Name |
| 12 | None | None | None | None | Simple | None | No Name |
| 13 | None | Recipe 1 | None | Coupon 1 | Simple | None | Sun Enlightened |
| 14 | ad 2 | None | None | Coupon 2 | Simple | None | BAD NAME |
| 15 | None | None | None | Coupon 1 | Complex | None | Sun Enlightened |
| 16 | None | Recipe 2 | None | None | Simple | Shopping Cart Ad 2 | Sun Enriched |
| 17 | None | None | None | None | Moderate | Shopping Cart Ad 1 | Sun Enriched |
| 18 | None | None | None | Coupon 1 | Simple | Shopping Cart Ad 2 | BAD NAME |
| 19 | ad 1 | None | Pop 2 | None | Simple | None | Sun Enlightened |
| 20 | None | None | Pop 2 | Coupon 2 | Simple | None | Sun Enriched |
| 21 | None | Recipe 1 | Pop 1 | Coupon 2 | Moderate | None | Sun Enriched |
| 22 | ad 1 | Recipe 1 | None | None | Moderate | Shopping Cart Ad 2 | No Name |
| 23 | ad 2 | Recipe 2 | Pop 1 | None | Moderate | None | Sun Enlightened |
| 24 | None | Recipe 1 | None | None | Complex | None | No Name |

*(continued)*

**TABLE 3.1 (continued)**

| Card ID | Floor Graphic | Recipe Holder | Shelf Pop-out | Coupon | Signage Levels | Shopping Cart | Vitamin D Name |
|---|---|---|---|---|---|---|---|
| 25 | ad 2 | Recipe 1 | Pop 2 | Coupon 1 | Complex | Shopping Cart Ad 2 | Sun Enriched |
| 26 | ad 1 | None | Pop 1 | None | Complex | Shopping Cart Ad 2 | BAD NAME |
| 27 | ad 2 | None | None | Coupon 2 | Moderate | Shopping Cart Ad 2 | Sun Enlightened |
| 28 | None | Recipe 2 | Pop 2 | None | Complex | Shopping Cart Ad 1 | Sun Enlightened |
| 29 | ad 1 | Recipe 1 | None | Coupon 2 | Simple | Shopping Cart Ad 1 | Sun Enlightened |
| 30 | None | None | Pop 1 | None | Simple | Shopping Cart Ad 2 | Sun Enlightened |
| 31 | ad 2 | Recipe 2 | Pop 1 | Coupon 1 | Simple | None | No Name |
| 32 | None | Recipe 1 | Pop 1 | None | Simple | None | BAD NAME |

Photoshopped for one card, resulting in 128 altered images for this design. The pictures were then placed into a Prezi document to simulate the flow of a shopper entering into the product section. A Prezi was made for each profile. The Prezi was then recorded via QuickTime, and the (mpv.) videos were embedded into the qualtrics survey platform. The videos were also placed on YouTube if the respondents didn't have compatibility with QuickTime. The YouTube links were clearly stated and placed underneath the embedded videos. Each video's length was one minute and five seconds.

## PROCEDURE

A total of 1,446 respondents were recruited through Amazon Mechanical Turk a crowdsourcing platform for human tasks. Two identical surveys were sent to males and females, the survey represented: 80.6% female (1,364 female and 328 male respondents) and 19.4% male observation rate. This rate was chosen because according to POPAI (The global association for marketing at retail) a 2012 study on shopper engagement concluded that 75% of shoppers are women. Women shoppers are dominating the aisles were men are just 25%. Additionally, women pay more attention to nutrition than men (Sanlier & Karakus, 2010) and due to particular marketing claims being represented, there was more emphasis placed on the female consumer. Furthermore, the age demographic chosen was 21 to 65, which consisted of 80% of the population that were primary food shoppers. The pool of potential respondents were restricted by location to be residents of the United States. The design of the study required them to open a link for the survey hosted on a qualtrics survey platform and then transfer their same credentials entered on qualtrics back to the MechianicalTurk webpage. Each respondent received a payment of $1.50.

The average survey duration was 9 minutes and 45 seconds. There were 1,159 women and 287 men, for a total of 1,446 respondents, who took the survey. The survey consisted of screening questions to ensure respondents' qualifications.

## RESULTS

Two different types of analysis were conducted in order to do a cross-validity examination. Both types of analysis were conducted for every demographic of consumers. The first was a stepwise regression of profile ratings, as the dependent variable (Y). The analysis was conducted in a stepwise approach to remove insignificant variables (Table 3.2). This is done because parameter estimates of the beta and intercept will change when a parameter is

**TABLE 3.2**

| Metric Regression Output | Beta Coefficients | P-Value |
|---|---|---|
| Intercept | 6.19 | <0.0001 |
| Age | 0.016 | 0.0021 |
| Income ($50,000–75,000) | 0.18 | 0.0051 |
| Complex Signage | 0.29 | <0.0001 |
| Shopping Cart Ad. # 1 | 0.21 | 0.002 |
| Shopping Cart Ad. # 2 | 0.368 | <0.0001 |
| R-Square | 4% | <0.0001 |

eliminated. There are five significant parameters in the model. The model is significant with an ANOVA p-value of <0.0001 and an adjusted R-square value of 0.039, meaning that the model accounts for 4% of the variability in consumers' attraction to the section.

The intercept is a significant parameter with a p-value of <0.0001 and a beta coefficient of 6.19. Age is a significant parameter with a p-value of 0.0021 and a beta coefficient of 0.016. Respondents who have an income between $50,000 and $75,000 are a significant parameter, with a p-value of 0.0051 and a beta coefficient of 0.18. Promotional signage that is complex is a significant parameter with a p-value of <0.0001 and a beta coefficient of 0.29. Promotional Shopping Cart Advertisement 1 is a significant parameter with a p-value of 0.002 and a beta coefficient of 0.21. Promotional Shopping Cart Advertisement 2 is a significant parameter with a p-value of <0.0001 and a beta coefficient of 0.368. The equations model is as follows: These results show that the starting intercept is 6.19, meaning that if there are no other factors to consider, the rating of the profile is 6.19. For every increase in a respondent's year of age, multiply by 0.016 in the rating of the profile. If a profile contains complex signage it will get a one-time bump of 0.29 in the profile rating. If a profile contains the shopping card Ad 1, it will get a one-time bump of 0.21 in the rating of the profile. If a profile contains the shopping card Ad 2, it will get a one-time bump of 0.36 in the rating of the profile.

The second analysis conducted was a logistic regression (Table 3.3). This is a choice-based analysis indicating the percent chance that the respondent chooses 1, the profile with additional stimuli, or –1, the existing section. The decimal values as coefficients represent the percent differences toward the respected decision to decide for the profile. The whole model test shows that the logistic regression is a significant model with a p-value of <0.0001. The R-square value is .038, which states that the independent variables explain 3.8% of the variation in the dependent variable. However, when doing a logistic regression with JMP, it doesn't automatically remove

**TABLE 3.3**

| Logit. Regression Output | Beta Coefficients | P-Value |
|---|---|---|
| Intercept | –0.59 | 0.0015 |
| Age | –0.012 | 0.0236 |
| Shelf Pop-out # 2 | 0.211 | 0.001 |
| Shopping Cart Ad. # 2 | 0.23 | 0.004 |
| Complex Signage | 0.42 | <0.0001 |
| R-Square | 4% | <0.0001 |

insignificant variables in a stepwise pattern. Once again, insignificant variables must be removed from the model because it affects the beta and significance levels of other variables. This process is done one at a time, removing the most insignificant values first. Variables are rejected from the research due to the protected error rate of the conjoint model; variables must have a p-value of 0.0032 or lower in order to be significant in the analysis. However, other independent variables such as age do not apply to the same standards due to not being a part of the conjoint design.

The logistic output shows a model with a p-value of 0.0015. Age was indicated as a significant variable with a p-value of 0.0236 and a beta coefficient of –0.012. Promotional Pop-Out Advertisement 2 was indicated as a significant parameter with a p-value of 0.001 and a beta coefficient of 0.211. Promotional Shopping Cart Advertisement 2 was indicated as a significant parameter with a p-value 0.0004 and the beta coefficient of 0.23. The last significant variable is complex signage in the section with a p-value of <0.0001 and a beta coefficient of 0.42. This equation shows that without any stimuli, respondents would be 59% more likely to choose the standard section. For every additional year in age, the respondent will be 1.2% less likely to choose the profile. If Promotional Pop-Out Advertisement #2 is present, then the respondent will be 21% more likely to choose the profile. If promotional Shopping Cart Advertisement #2 were present, then the respondent would be 23% more likely to choose the profile. Finally, if there were complex signage present, then the respondent would be 42% more likely to choose the profile. This data indicate a heavy preference for the standard section, although opinions were altered when shown the advertisements listed above. The most effective promotional material was complex signage. Second was the shopping cart advertisement, followed by the promotional pop-out ad. Interaction effects were conducted among the significant promotional parameters; no interaction effect was found for either analysis.

## Analysis by Gender

Only the female gender yielded significant output (see Table 3.4). The first analysis (metric) with rating as the dependent variable developed a significant model with an ANOVA p-value of <0.0001, and a R-square value of 3%. Indicating that only 3% of the variability in ratings is explained by the independent variables. The first independent variable is the intercept, with a beta coefficient of 6.658. The intercept indicates that if no other independent variable is present, then the rating respondents would give would be 6.658. The second independent variable, complex signage, has a beta coefficient 0.33 with a p-value of <0.0001, indicating that if complex signage is present within a profile, then the profile will be 0.33 points higher on the attractiveness Likert scale. Shopping Cart Ad 1 is the next significant variable with a beta coefficient of 0.242 Shopping Cart Ad 1 with a P-value of <0.0001. This indicates that if Shopping Cart Advertisement 1 is present in a profile, then it will be rated 0.242 higher on the attractiveness scale. Shopping Cart Ad 2 is the next significant variable with a beta coefficient of 0.242 and a P-value of <0.0001. This indicates that if Shopping Cart Advertisement 1 is present in a profile, then it will be rated 0.357 higher on the attractiveness scale.

The next analysis for gender is the non-metric logistic regression, which represents the choice based results. The model was significant with p-value of <0.0001 and an R-square of 4%. Only 4% of the variability in respondents choosing the profile or the standard can be explained by the independent variables. The independent variables are: Intercept with a beta coefficient of –1.02, which indicates that a respondent is 100% likely to pick the standard if there is no other independent variables present. If the next significant variable, Shelf Pop-out 2, is present, then respondents would be 25%

**TABLE 3.4**

| Metric Regression Output | Beta Coefficients | P-Value |
|---|---|---|
| Intercept | 6.658 | <0.0001 |
| Complex Signage | 0.33 | <0.0001 |
| Shopping Cart Ad. # 1 | 0.242 | <0.0001 |
| Shopping Cart Ad. # 2 | 0.357 | <0.0001 |
| R-Square | 3% | <0.0001 |
| **Logit. Regression Output** | | |
| Intercept | –1.02 | <0.0001 |
| Shelf Pop-out # 2 | 0.253 | 0.0005 |
| Complex Signage | 0.488 | <0.0001 |
| Shopping Cart Ad. # 2 | 0.211 | 0.0045 |
| R- Square | 4% | <0.0001 |

more likely to choose the profile over the standard. The next significant variable, complex signage, has a beta coefficient of 0.488 and a p-value of <0.0001, indicating that if complex signage is present in the profile, then the respondent is 49% more likely to pick the profile over the standard. Shopping Cart Advertisement 2 is the next significant variable with a beta coefficient of 0.211 and a p-value 0.0045.

## Analysis by Age

The first analysis done for the respondents in their 20s was the logistic regression, where Y is the choice-based (non-metric) decision between the standard and the profile (see Table 3.5). The model was significant, with a p-value of <0.0001 and an R-square 3%, indicating that only 3% of the variability in respondents choosing the standard or profile can be explained by independent variables in this model. The independent variable was the intercept, which has a beta coefficient of –0.81, indicating that with no other independent variables present, respondents would be 81% more likely to choose the standard over the profile. The next significant variable is complex signage, with a beta coefficient of 0.42 and a p-value of <0.0001, indicating that if complex signage is present within the profile, respondents are 42% more likely to choose the profile over the standard. The next significant variable, Shopping Cart 2, had a beta coefficient of 0.28 and a p-value of 0.0025, indicating that if Shopping Cart Ad 2 is present in the profile, then the respondent would be 28% more likely to pick the profile over the standard.

The second analysis conducted for the respondents in their 20s is the metric regression. Where the dependent variable is the rating of attractiveness. The model was significant, with an ANOVA P-Value of <0.0001 and a beta coefficient and an R-square of 3%, which states that only 3% of the variability of in the rating of attractiveness can be explained by the

**TABLE 3.5 Respondents in the 20s**

| Logit. Regression Output | Beta Coefficient | P-Value |
|---|---|---|
| Intercept | –0.81 | <0.0001 |
| Complex Signage | 0.42 | <0.0001 |
| Shopping Cart #2 | 0.28 | 0.0025 |
| R-Square | 3% | <0.0001 |
| **Metric Regression Output** | | |
| Intercept | 6.17 | <0.0001 |
| Complex Signage | 0.34 | <0.0001 |
| Floor Graphic # 2 | –0.30 | 0.00119 |
| R-Square | 3% | <0.0001 |

**TABLE 3.6 Respondents in the 30s**

| Logit. Regression Output | Beta Coefficient | P-Value |
|---|---|---|
| Intercept | –0.81 | <0.0001 |
| Complex Signage | 0.49 | <0.0001 |
| R-Square | 6% | <0.0001 |

independent variables. The independent variable is the intercept 6.17, indicating the average rating respondents in their 20s give to the profile without adjusting for the other independent variables. The next significant variable is complex signage, with a beta coefficient of 0.34 and a p-value of <0.0001, indicating that the presence of complex signage in the profile will increase the profile's rating by 0.34. The next significant variable is Floor Graphic 2, with a beta coefficient of –0.3 and a p-value of 0.00119, indicating that if a profile has the presence of Floor Graphic 2, then respondents in their 20s are going to rate the profile –0.3 less on the rating scale.

The only analysis that was significant with respondents in their 30s was the non-metric model (Table 3.6), where the dependent variable was the choice-based decision of respondent choosing either the standard or the profile. The model had an R-square of 6% and a p-value of <0.0001, which states only 6% of the variability of respondents' choices can be explained by the independent variables. The independent variable, the intercept, has a beta coefficient of –0.81, indicating that with no other independent variables present, the respondents in their 30s are 81% more likely to choose the standard over the profile. The only other significant variable is complex signage, with a beta coefficient of 0.49 and a p-value of <0.0001, indicating that when complex signage is present within the profile, respondents in their 30s are 49% more likely to choice the profile over the standard. No significant model existed for rating the metric dependent variable, and there were no significant outputs for any of the existing age groups.

## Analysis by Income

For respondents with income of less than $25,000, only the logistic (non-metric) analysis was a significant model (see Table 3.7). The logistic model

**TABLE 3.7 Respondents with an Income of Less than $25,000**

| Logit. Regression Output | Beta Coefficients | P-value |
|---|---|---|
| Intercept | –0.89 | <0.0001 |
| Complex Signage | 0.591 | 0.0008 |
| R-Square | 4% | 0.0003 |

has an R-square of 4% and a p-value of 0.0003, stating that only 4% of the variability of respondents with an income of less than $25k can be explained by the independent variables. The independent variable, intercept –0.89, indicates that with no other independent variables present, respondents earning less than $25k are 89% more likely to choose the standard than the profile. The only other significant variable is the complex signage, with a beta coefficient of 0.59 and a p-value of 0.0008, indicating that with the presence of complex signage in the profile, respondents with an income less than $25k are 59% more likely to pick the profile over the standard. No significant results were present for rating as a dependent variable.

The first analysis for respondents with an income of $25,000 to $49,999 (Table 3.8) is the metric regression model where the dependent variable is the rating of attractiveness. The model has an ANOVA p-value of <0.0001 and an R-square of 3%, indicating that only 3% of the variability of respondents with income between $25k and $50k in the rating of attractiveness can be explained by the independent variables. The independent variables, intercept 6.52, indicates that no if other independent variables are present, the profile would receive a rating of 6.52. The only other significant variable is complex signage, with a beta coefficient of 0.48 and a p-value of <0.0001, indicating that if the profile has complex signage, then respondent whose income is between $25k and $50k are going to rate the attractiveness of the profile 0.48 points higher.

The second analysis conducted for respondents with an income of $25,000 to $49,999 is the logistic regression model (non-metric), where the dependent variable is the respondents' choice of either the standard or the profile. The model's p-value was <0.0001 with an R-square of 5%, indicating that only 5% of the variability in respondents choosing either the standard or the profile can be explained by the independent variables. The independent variables are the intercept with a beta coefficient of –0.81, which indicates that if no other independent variables exist in the profile, then respondents whose salary is between $50k and $75k are 81% more likely to

**TABLE 3.8 Respondents with an income of $25,000 to $49,999**

| Metric Regression Output | Beta Coefficients | P-value |
|---|---|---|
| Intercept | 6.52 | <0.0001 |
| Complex Signage | 0.48 | <0.0001 |
| R-Square | 3% | <0.0001 |
| **Logit. Regression Output** | | |
| Intercept | –0.811 | <0.0001 |
| Complex Signage | 0.671 | <0.0001 |
| R-Square | 5% | <0.0001 |

**TABLE 3.9 Respondents with an Income of $50,000 to $74,999**

| Metric Regression Output | Beta Coefficients | P-value |
|---|---|---|
| Intercept | 6.735 | <0.0001 |
| Shopping Cart Ad # 2 | 0.58 | <0.0001 |
| R-Square | 5% | <0.0001 |

choose the standard over the profile (Table 3.9). The only other significant variable is complex signage, with a beta coefficient of 0.671 and a p-value of <0.0001, indicating that the presence of complex signage in the profile would cause respondents whose income is between $50k and $75k to be 67% more likely to choose the profile.

The first analysis for respondents with an income of $50,000 to $74,999 is a metric regression where the dependent variable was the rating of attractiveness. The model's ANOVA p-value was <0.0001 with an R-square of 5%, indicating that only 5% of the variability in the rating of attractiveness by respondents with incomes between $50k and $75k can be explained by the independent variables (intercept: 6.74), indicating that if there is no other presence of independent variables from the model, then the respondents with an income between $50k and $75k would rate the profile 6.74. The only other significant variable is Shopping Cart Advertisement 2, with a beta coefficient of 0.58 and a p-value of <0.0001, indicating that when the presence of Shopping Cart Advertisement 2 was in a profile, respondents whose income was between $50k and $75k would rate the profile 0.58 points higher. There was no significant output for the logistic regression.

The first analysis for respondents who had an income of $75,000 to $99,999 was a metric regression where the dependent variable was the rating of attractiveness (Table 3.10). The model's p-value was 0.0028 and it had an R-square of 5%, indicating that only 5% of the variability in the ratings of attractiveness by respondents with an income between $75k and $100k can be explained by the independent variables. The independent variable, the intercept 6.26, indicate that with no other independent variables present in the profile, respondents who earn between $75k and $100k would rate the profile 6.3. The only other significant variable is Shopping Cart

**TABLE 3.10 Respondents with an Income of $75,000 to $99,999**

| Metric Regression Output | Beta Coefficients | P-Value |
|---|---|---|
| Intercept | 6.263 | <0.0001 |
| Shopping Cart Ad # 2 | 0.541 | 0.0028 |
| R-Square | 5% | 0.0028 |

Advertisement 2, with a p-value of 0.0028 and a beta coefficient of 0.541, indicating that when Shopping Cart Advertisement 2 is present within a profile, respondents with an income of $75k to $100k will rate the profile 0.541 points higher. There was no significant output for logistic regression. All other income levels had no significant outputs for either analysis.

## Analysis by Ethnicity

There was only one demographic of race that yielded significant outputs, Caucasians. Two analyses were run for Caucasians. The first was the metric regression where the dependent variable was the rating of attractiveness. The model's ANOVA p-value was <0.0001 with an R-square of 3%, indicating that only 3% of the variability in the rating of attractiveness by Caucasians could be explained by the independent variables (see Table 3.11). The independent variable, the intercept, had a beta coefficient of 6.6, indicating that if no other independent variables were present from the model in the profile, Caucasians would rate the profile 6.6. The next significant variable is complex signage, with a beta coefficient of 0.34 and a p-value of <0.0001, indicating that if the presence of complex signage existed in a profile, then the Caucasian respondents would rate the profile 0.34 points higher. The second significant variable was Shopping Cart Advertisement 1 with a beta coefficient of 0.27 and a p-value of 0.002, indicating that the presence of Shopping Cart Advertisement 1 in a profile would cause Caucasian respondents to rate the profile 0.27 points higher. The last significant variable in the model is Shopping Cart Advertisement 2 with a beta coefficient of 0.34 and a p-value of 0.0002, indicating that the presence of Shopping Cart

**TABLE 3.11**

| Metric Regression Output | Beta Coefficients | P-Value |
|---|---|---|
| Intercept | 6.604 | <0.0001 |
| Complex Signage | 0.34 | <0.0001 |
| Shopping Cart Ad # 1 | 0.27 | 0.002 |
| Shopping Cart Ad # 2 | 0.34 | 0.0002 |
| R-Square | 3% | <0.0001 |
| **Logit. Regression Output** | | |
| Intercept | –1.206 | <0.0001 |
| Recipe Holder # 1 | 0.23 | 0.0076 |
| Shelf Pop-out # 2 | 0.28 | 0.0006 |
| Shopping Cart Ad. # 2 | 0.29 | 0.0008 |

Advertisement 2 would result in Caucasian respondents rating the profile 0.34 points higher.

The second analysis run for a Caucasian respondent was the non-metric logistic regression where the dependent variable was the choice Caucasians picked—either the standard or the profile (see Table 3.12). The model's p-value was <0.0001 with an R-square of 4%, indicating that only 4% of the variability in Caucasian respondents choosing either the standard or profile could be explained by the independent variables. The independent variable, the intercept with a beta coefficient of –1.20, means that Caucasian respondents are 121% more likely to pick the standard, if no other independent variables are present from the model. The next significant variable is Recipe Holder 1 with a p-value of 0.0076 and a beta coefficient of 0.23, indicating that when the profile has the presence of Recipe Holder 1, Caucasian respondents are 23% more likely to choose the profile. The next significant variable was Shelf Pop-Out 2 with a p-value of 0.0006 and a beta coefficient of 0.28, indicating that the presence of Shelf Pop-Out #2 in a profile will make Caucasian respondents 28% more likely to pick the profile. The last significant variable was Shopping Cart Advertisement 2 with a beta coefficient of 0.29 and a p-value of 0.0008, indicating that if the presence of Shopping Cart Ad 2 existed in the profile, Caucasian respondents would be 29% more likely to pick the profile.

While the average rating of the profiles was not in the design, it is interesting to note that the top three rated profiles (profile 28, profile 11, and profile 28) all have complex signage embedded.

**TABLE 3.12 Top Rated Profiles**

| Profile | Rating Score | Profile | Rating Score |
|---|---|---|---|
| 1 | 6.52 | 17 | 5.47 |
| 2 | 6.56 | 18 | 6.36 |
| 3 | 5.39 | 19 | 5.37 |
| 4 | 6.11 | 20 | 5.90 |
| 5 | 6.23 | 21 | 6.15 |
| 6 | 5.94 | 22 | 6.30 |
| 7 | 6.20 | 23 | 5.74 |
| 8 | 6.06 | 24 | 6.65 |
| 9 | 5.64 | 25 | 7.08 |
| 10 | 5.88 | 26 | 6.29 |
| 11 | 7.06 | 27 | 6.00 |
| 12 | 5.61 | 28 | 6.71 |
| 13 | 5.33 | 29 | 5.85 |
| 14 | 5.30 | 30 | 6.57 |
| 15 | 6.00 | 31 | 5.55 |
| 16 | 6.45 | 32 | 5.83 |

## CONCLUSION

Instituting promotional stimuli can be a massive investment; due to the mushroom industry not having the financial capital of CPG companies, a more segmented analysis was conducted. In reality the mushroom industry would not roll out these promotions nationwide but rather by individual accounts or by independent grocery retailers. These accounts and stores may have varying clientele. By having a deeper understanding of what affects each demographic, these results can have a more effective outcome.

Women in both analyses showed that complex signage was the most effective stimulus in attracting customers, along with Shopping Cart Advertisement 2. These two findings reiterate the overall analysis (most likely due to women being 80% of the survey population), but an interesting finding is that Shelf Pop-Out 2 attracted women respondents more than Shopping Cart Advertisement 2. The shelf pop-out ended up being half as effective as the complex signage.

Respondents in their 20s had very similar attraction levels to complex signage compared to the overall analysis. The attractiveness levels of shopping cart advertisements decreased and, most importantly, the presences of Floor Graphic 2 had a negative effect. Thus, if a specific supermarket has a younger shopper demographic it would behoove the mushroom industry to put the majority of the investment into complex signage while allocating none of the budget to the floor graphics. Floor graphics carried a negative sentiment from the consumer. The next age group, respondents in their 30s, showed a higher level of attractiveness to the complex signage. This could be a potential segmentation strategy to maximize returns on promotional investment.

Income analysis has shown that respondents who earn less than $25,000 are most affected by complex signage. This could be because respondents at a lower economic status might not be accustomed to any promotional signage whatsoever. This result may be due to these respondents shopping at lower-end supermarkets that just focus on lowering prices, and not on promotions. Additionally, these consumers could very well be located in a food desert and not have access to a supermarket. Instituting complex signage would definitely be successful in attracting them to the aisle; however, they most likely are not going to buy mushrooms due to their limited budgets.

Respondents with an income of $25,000 to $49,999 show that complex signage attracts this income bracket higher than any other income bracket. The mushroom industry instituted complex signage with supermarkets that have a large portion of their population in the income ranges of $25,000 to $50,000. As incomes increase, the effectiveness of complex signage becomes insignificant, whereas for any income level above $50,000, respondents showed a high attraction to Shopping Cart Advertisement 2. An analysis of an additional demographic of ethnicity was conducted, but there was only one significant model that arose. Therefore, it would be redundant to form any conclusions based on the output from the model.

## REFERENCES

Abratt, R., & Goodey, S. (1990). Unplanned buying and in-store stimuli in supermarkets. *Managerial and Decision Economics, 11*(2), 111–121.

AC-Nielsen Company. (2006). *Consumer-centric category management: How to increase profits by managing categories based on consumer needs.* Hoboken, NJ: Wiley

Advertising Age. (2007, June 25). *100 leading national advertisers.*

Ailawadi, K. L., Beauchamp, J. P., Donthu, N., Gauri, D., & Shankar, V. (2009). Communication and promotion decisions in retailing: A review and directions for future research. *Journal of Retailing, 85*(1), 42–55.

Bosselmann, P., & Kenneth, C. (1987). Perceptual simulations of environments. In R. B. Bechtel et al. (Eds.), *Methods in Environmental and Behavioral research* (pp. 162–190). New York, NY: Van Nostrand Reinhold.

Kiran, V., Majumdar, M., & Kishore, K. (2012). Innovation in in-store promotions: Effects on consumer purchase decision. *European Journal of Business Management, 4*(9), 36–44.

Linton, I., & Demand Media. (n.d.). What is the failure rate of new items launched in the grocery industry? *Chron.* Retrieved from http://smallbusiness.chron.com/failure-rate-new-items-launched-grocery-industry-13112.html

Lite, J. (2009). Vitamin D deficiency soars in the U.S. *Scientific American.* Retrieved from http://www.scientificamerican.com/article/vitamin-d-deficiency-united-states/

Livestock Marketing Association (LMA) v. USDA (132 F. Supp. 2d 817 (D. S.D. 2001), and Johanns, et al. v. LMA (544 U.S. 550 2005)

McKechnie, G. E. (1977). Simulation of techniques in environmental psychology. In D. Stokols (Ed.), *Perspectives on environment and behavior: Theory, research and applications* (pp. 169–189). New York, NY: Plenum.

Olver, J. M., & Farris, P. W. (1989). Push and pull: A one-two punch for packaged goods. *Sloan Management Review, 31*(1), 53–61.

Sanlier, N., & Seren Karabus, S. (2010). Evaluation of food purchasing behavior of consumers from supermarkets. *British Food Journal, 112*(2), 140–150.

Steel, E. (2008, August 21). The ad changes with the shopper in front of it. *Wall Street Journal,* page B8.

USDA. (2013, February 6). *Economic research service processing & marketing: New products.* Retrieved from http://www.ers.usda.gov/media/1037954/eib_108

CHAPTER 4

# METHODS FOR CUSTOMER ANALYTICS OF HETERGENEOUS E-COMMERCE POPULATIONS

**Ruben Mancha**
***Babson College***

**Mark T. Leung**
***University of Texas at San Antonio***

## ABSTRACT

The generation of business intelligence from the analysis of customer data has become an imperative issue in contemporary business management. One of the key areas in customer analytics is how to effectively obtain insight from consumer behavior and characteristics through data mining. In this chapter, we describe three statistical methods for the segmentation analysis of a heterogeneous population of customers. All three methods are variations of the general partial least squares path modeling (PLS-PM) approach for structural equation modeling (SEM). The first method, hierarchical clustering of PLS scores (HC-PLS), employs general PLS-PM procedures attending to a preconceived theoretical model and uses the estimated PLS scores of each observation to guide hierarchical clustering (HC) and identify customer segments. The sec-

*Contemporary Perspectives in Data Mining, Volume 2,* pages 55–75

ond method is finite mixture partial least squares (FIMIX-PLS), an integrated approach that segments the heterogeneous population and assigns units to respective segments, while accounting for the measurement and inner path model estimates. The third one, response-based unit segmentation partial least squares (REBUS-PLS), identifies the distinctive segments and performs PLS-PM on each one of them, leading to segments that are different in both measurement and inner models. All three methods have shown in the literature some degree of success in dealing with unobserved heterogeneity inherent to customer data. For illustrative purposes, our chapter also contains an empirical analysis of online auction customer data using these three methods, along with a hierarchical clustering benchmark. Results indicate that different methods may lead to dissimilar segmentation schemes and varied conclusions, suggesting that further research is needed to develop a unified metric for comparison of the efficacy of these methods and to explore their limitations.

## INTRODUCTION

Rapid adoption of computer and information technology in the last decades has allowed market researchers to evaluate customers like never before, analyzing the sheer volume of complex data, capturing customers' profiles and behavior. Such customer analytics often lead to business intelligence guiding the formulation of product and service strategies, as well as refinement of operational plans. Hence, performing customer analytics has become an imperative issue in contemporary management practice. One of the key areas in customer analytics is how to effectively extract accurate information of consumer behavior and characteristics through data mining.

Many studies in customer analytics assume homogeneity of the customer base (e.g., Lyk-Jensen & Chanel, 2007). The identification of customer types is becoming a priority among business intelligence providers (e.g., IBM and SAP) and for online businesses with increasing amounts of data about their customers. In this chapter, we demonstrate, without *a priori* knowledge on the number and size of segments and attending to a theoretical model, how to empirically identify heterogeneous subgroups (segments) in a population of customers and classify units (customers) with similar characteristics into these segments. The methods we describe mirror the real-world situation faced by market researchers—they have large amounts of customer data with myriad attributes and exhibited behaviors but do not know how many customer *types* these profiles and behaviors represent. Furthermore, even previously identified customer segments can be fluid and change over time. Hence, it is useful to employ methodologies that do not require stringent assumptions and concrete *a priori* knowledge of customer segments.

In the past, clustering, discriminant analysis, and artificial neural networks have been applied to the problem of market segmentation (Kuo, Ho,

& Hu, 2002). These methods' requirements vary from *a priori* identification of segments to data preprocessing and preclassification. In particular, these classification methods fail at identifying or capturing possible causal relationships among environmental variables, individual traits, and/or human behavior. Moreover, clustering and artificial neural network methods, even efficient in finding relevant segments, are not guided by theoretical constructs, nor can they offer researchers meaningful insights about the consumer's decision process.

In this chapter, we describe three statistical methods for segmentation analysis of a heterogeneous population of online customers. All three methods are extended variations of the general partial least squares path modeling (PLS-PM) approach for structural equation modeling (SEM). The first method is hierarchical clustering of PLS scores (HC-PLS). Guided by a preconceived theoretical model, this method employs the general PLS-PM procedure to the entire dataset and then uses the estimated PLS scores of each observation to perform hierarchical clustering (HC) for identification of customer segments. The second method is finite mixture partial least squares (FIMIX-PLS), an integrated approach that segments the heterogeneous population and assigns units to segments, while accounting for the path model estimates. The third one, response-based unit segmentation partial least squares (REBUS-PLS), is a more recently developed method to simultaneously execute PLS-PM and identify distinctive segments in the data through hierarchical clustering on the computed residuals of the measurement and inner models. In the literature, all three methods have shown some degree of success in dealing with unobserved heterogeneity inherent to customer datasets. For illustrative purposes, our chapter demonstrates an empirical analysis of an online auction dataset using these three methods. To create a baseline for benchmarking, a statistical analysis utilizing the traditional hierarchical clustering (HC) approach, outside the context of PLS-PM, is also included. It is important to point out that the traditional HC method is a naïve approach and does not take into consideration the causal relationships uncovered in the PLS-PM model. Segmentation by the HC method simply treats the variables as independent inputs to the multivariate analysis.

The rest of the chapter is organized as follows: Section 2 describes the three segmentation methods used in conjunction with PLS modeling. The section also briefly summarizes the traditional (non-PLS-based) clustering method, which establishes a baseline for comparison. Section 3 describes the illustrative empirical analyses of customer data collected from an online auction experiment. Results from the three PLS segmentation methods reviewed in this study along with those from the baseline method are reported and discussed. Section 4 concludes the chapter and offers remarks and suggestions for future research.

## METHODS

### Partial Least Squares Path Modeling (PLS-PM)

Partial least squares path modeling (PLS-PM) is one of the methodologies used for estimation of structural equation models (SEM). It has been gaining popularity after the seminal work by Wold (1975a, 1975b), especially among research studies in the fields of social sciences and business. Some of its major business applications in recent years are related to consumer analytics and to the examination of customer characteristics and behaviors in e-commerce.

In simple terms, SEM is used to establish a system of causal relationships among different entities in a theoretical model. These entities, called latent variables, are often not directly or completely observable. Thus, each latent variable is, in turn, indirectly established (or "proxied") by a set of measurement variables, also known as manifest variables. Hence, the main purpose of PLS-PM is to combine blocks of observed indicators into latent variables and then estimate the causality links among the latent variables according a prespecified theoretical and hypothesized model. As one can visualize, the procedure of PLS-PM embraces statistical estimations of outer (measurement) and inner (structural) models through iterative steps. After the estimation is completed, the entire PLS model is then evaluated by various criteria for its empirical suitability for the study or effectiveness (see Sarstedt, Becker, Ringle, & Schwaiger, 2011, for additional discussion on the topic). Literature regarding PLS-PM and its technical exposition is abundant and extensively documented. For the sake of brevity, interested readers can refer to Tenenhaus, Esposito Vinzi, Chatelin, and Lauro (2005) and Esposito Vinzi, Trinchrea, and Silvano, (2010) for detailed discussions on the topic.

### Hierarchical Clustering of Partial Least Squares Scores (HC-PLS)

Estimation by PLS-PM assumes the presence of a homogenous population; that is, the observed indicators and thereby model variables come from a group exhibiting similar relationships among the studied characteristics and behaviors. This assumption may not be met in many common situations. For instance, it is unreasonable to expect that few latent variables about online customers' preferences and behaviors relate to each other independently from unaccounted variables. Consequently, the estimated PLS-PM coefficients and their significance can be distorted, leading to wrongful guidance from the results. This accentuates in the early stages of theoretical development, when the models are unrefined and the presence of moderating effects is yet to be uncovered.

For example, as we empirically evaluate in Section 3, the study of online auctions has been concerned with those factors leading to increased willingness to pay. While some shoppers are proactive or strategic in making purchasing decisions and seek to minimize payment and effort, others are compelled by convenience or enjoyment derived from the shopping experience and are willing to pay extra for additional services or thrilling purchasing experiences. A theoretical model evaluating how the presence of these service features affect final customer payments, not accounting for the level of hedonic consumption and simply estimated by the sole means of PLS-PM, would erroneously generate path models reconciling the opposing relationships between the latent variables if the significant variation among customers is not considered.

A direct remedy to this adverse effect is to consider the population as multiple heterogeneous groups and apply clustering procedures to identify and create subgroups (or segments). This method explicitly mandates the need of segmentation of data (i.e., classification of observations) after PLS-PM analysis. In short, the HC-PLS method employs a general PLS-PM procedure on the entire heterogeneous dataset and then, instead of calculating the aggregated path coefficients, uses the estimated PLS scores for each observation to guide hierarchical clustering (HC) and identify customer segments. As a result, the units (or customers) lumped in a cluster should share commonalities in terms of relationships between the latent variables. This method also allows the researcher to compare the similarities and contrast the differences across various segments (local models) and with respect to the entire population. The evaluation of each segment with PLS-PM should permit the identification of differences across path coefficients, and multigroup comparison methods such as permutation procedures offer statistical tests aiming at comparison of segments. It should be noted that HC-PLS is a two-stage sequential method. Hence, it may embrace a higher level of potential error than those integrated methods that consider segmentation and PLS-PM simultaneously. In addition, unless guided by established work, the researcher faces the challenge of determining the value of $K$ (the number of segments) prior to the empirical analysis.

Generally speaking, HC involves the development of a hierarchy or tree structure for the purpose of segmenting the units in dataset. There are two ways to form the clusters in a hierarchical manner—agglomerative and divisive. The agglomerative approach treats each unit as a cluster initially. In subsequent steps, the two closest clusters are merged into a new aggregate cluster. Eventually, all units are combined together into one global cluster. The divisive approach operates in the opposite fashion. All units in the dataset belong initially to one cluster. In each following step, units that are most dissimilar are divided into separate clusters. The process continues until each unit represents its own cluster. HC is a well-established method and has been extensively documented in the literature. For instance, readers can refer to Hair, Anderson, Tatham, and Black (1998) for a technical exposition of HC along with its options and limitations.

No matter whether the agglomerative or divisive approach is adopted, an issue encountered in the HC process is the selection of the best number of clusters ($K$), which is unknown *a priori.* Although some criteria and tools exist (such as the dendrogram graphic), there is no general consensus as to the standard way to determine the number of segments $K$. Hair et al. (1998) recommend computing the solutions for different numbers of segments and then using prevailing theories or practical judgment to guide the selection of $K$.

## Finite Mixture Partial Least Squares (FIMIX-PLS)

Essentially an extension of partial least squares path modeling (PLS-PM), finite mixture partial least squares (FIMIX-PLS, or simply FIMIX) is the second method we will describe for the segmentation of a heterogeneous population of e-commerce customers. As stated earlier, a major assumption associated with PLS-PM is that the dataset originates from a single homogeneous population. Although this underlying assumption is met in some research scenarios, PLS-PM is not an appropriate method when examining a diverse pool of subjects across a spectrum of background characteristics. Failing to identify the possible heterogeneity and to account for segmentation in the global PLS-PM model may lead to inappropriate results and flawed conclusions (Sarstedt, Schwaiger, & Ringle, 2009; Ringle, Sarstedt, & Mooi, 2009). Hutchinson, Kamakura, and Lynch (2000) discussed the basic problems arising from unobserved heterogeneity in behavioral research and outlined suggestions for this issue. As a result, several approaches, including FIMIX-PLS, have been developed to capture the heterogeneity in unsegmented datasets. Sarstedt (2008) provides a survey of these approaches. The FIMIX-PLS approach allows model parameters to be estimated and subject affiliations to be simultaneously segmented (Sarstedt & Ringle, 2010). Furthermore, FIMIX-PLS considers both inner (structural) and outer (measurement) models during its computational process, whereas HC-PLS examines only the inner model for segmentation. Theoretically, FIMIX-PLS should yield a better fit than HC-PLS in terms of representing multiple segments within the global model.

FIMIX-PLS is a method of choice when there is neither general consensus as to the exact number of bidder types in online auctions, nor established literature with a structured list of attributes to segment/classify bidders. Besides, FIMIX-PLS can eliminate an additional layer of potential errors stemming from data pre-processing procedure—for example, applying *k*-mean clustering or discriminant analysis to preclassify the observations into disjoint bidder segments.

The FIMIX-PLS method specifically copes with ordinary least squares (OLS)-based predictions from the PLS model (Hahn, Johnson, Herrmann, & Huber, 2002). It combines finite mixture models with the expectation-maximization (EM) algorithm. Segmentation is performed in an iterative fashion over a range of segment sizes. Ringle et al. (2010) provided numerical examples.

For models with finite mixture, it is assumed that the dataset is derived from a segmented population comprised of $K$ subpopulations that can be modeled separately. Thus, the overall population is a mixture of these subpopulations where each observation $x_i$, for $i = 1, \ldots, N$, is drawn from a mixture density with $K$ segments. Following the notations from Sarstedt et al. (2009), this concept is summarized in the mixture density in Equation 4.1:

$$f_{i|k}(x_i|\theta_k) = \sum_{k=1}^{K} \rho_k f_{i|k}(x_i|\theta_k), \tag{4.1}$$

where $\rho_k$ is the relative size of segment $k$, $\rho_k > 0$,

$$\sum_{k=1}^{K} \rho_k = 1,$$

$f_{(i|k)}(\cdot)$ is the density function of segment $k$, and $\theta_k$ represents the segment-specific vector of unknown parameters for segment $k$. This mixture density captures the characteristics of each bidder segment and, on an aggregate basis, the bidder segments form the population of the online auction bidders examined in the study. The optimal number of segments $K$ (which is not observable *a priori*) is determined through examination of fit indices (e.g., Akaike Information Criterion and entropy statistic).

This approach intrinsically categorizes the bidders into $K$ segments with respect to the latent variables and measured variables as well as the modeled relationships between latent variables (see Figure 4.1). The initial FIMIX-PLS step is to estimate a path model using the PLS-PM algorithm on the manifest variables in the outer models (Sarstedt et al., 2009). The resulting latent variable scores are then employed to run the second step of the FIMIX-PLS algorithm in the inner model. Hahn et al. (2002) and Sarstedt et al. (2009) provided detailed technical expositions of the approach.

Based on the conceptual foundation outlined in Ringle, Wende, and Will (2005), each endogenous latent variable $\eta_i$ is distributed as a finite mixture of conditional density functions. This segment-specific conditional density function is expressed as

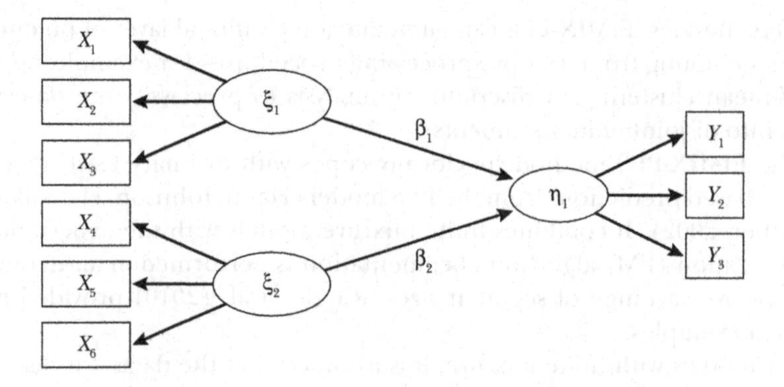


**Figure 4.1** PLS path model.

$$\eta_i \sim \sum_{k=1}^{K} \rho_k f_{i|k}(\eta_i | \xi_i, B_k, \Gamma_k, \Psi_k), \tag{4.2}$$

where $\xi_i$ is a vector of exogenous variables in the inner model in respect to observation $i$, $B_k$ is the path coefficient matrix of the endogenous variables, $\Gamma_k$ is the path coefficient matrix of exogenous latent variables, and $\Psi_k$ is the matrix of each segment's regression variances of the inner model. If each variable $\eta_i$ is assumed to have a multivariate normal distribution, the exact form of the density function can thus be written as

$$\eta_i \sim \sum_{k=1}^{K} \rho_k \left[ \frac{B_k}{2\pi^{M/2} |\Psi_k|^{1/2}} \exp\left( -\frac{1}{2} (B_k \eta_i + \Gamma_k \xi_i)' \Psi_k^{-1} (B_k \eta_i + \Gamma_k \xi_i) \right) \right], \tag{4.3}$$

where $M$ indicates the number of endogenous latent variables in the inner model. Model parameters are then estimated using maximum likelihood method with the log-likelihood function

$$LnL_c = \sum_{i=1}^{N} \sum_{k=1}^{K} z_{ik} \ln\left( f_{i|k}(\eta_i | \xi_i, B_k, \Gamma_k, \Psi_k) \right) + \sum_{i=1}^{N} \sum_{k=1}^{K} z_{ik} \ln(\rho_k), \tag{4.4}$$

where $z_{ik}$ is a binary variable indicating the assignment of observation $i$ to segment $k$. Observation $i$ is not assigned to segment $k$ if $z_{ik} = 0$ and is assigned to segment $k$ if $z_{ik} = 1$. Classification is dependent upon a combination of observed variables (behaviors, attitudes, traits, etc.) and hypothesized relationships between their underlying constructs.

The model parameters are estimated with a modified version of the EM algorithm, which allows for simultaneous and independent estimation of the segment-specific regression functions. Specifically, the EM algorithm finds the $K$ sets of maximum likelihood estimators using OLS (Hahn et al., 2002; Ringle et al., 2005a). Because the exact, or optimal, number of segments is unknown *a priori*, different values for $K$ within a reasonable range have to be evaluated iteratively. Based on the estimates, FIMIX-PLS can compute the expected conditional probability for observation $i$ in segment $k$ using Bayes' theorem:

$$P_{ik} = \frac{\rho_k f_{i|k}(\eta_i|\xi_i, B_k, \Gamma_k, \Psi_k)}{\sum_{k=1}^{K} \rho_k f_{i|k}(\eta_i|\xi_i, B_k, \Gamma_k, \Psi_k)}, \forall i, k = 1, \cdots, K. \quad (4.5)$$

Bidder $i$ is assigned to segment $k$ if

$$P_{ik} = \max\{P_{ik'}|k' = 1, \cdots K\}. \quad (4.6)$$

Pseudo code of the EM algorithm is presented in Ringle et al. (2005). The algorithm requires no *a priori* classification of bidders by type, although the number of segments $K$ must be designated.

The EM algorithm (i.e., the first step of our FIMIX-PLS framework) follows the standard PLS-PM modeling procedure (Sarstedt et al., 2009; Ringle et al., 2005). The loadings of the latent variables in the inner path model are then used for the FIMIX-PLS procedure.

The determination of $K$ is based on a set of criteria that measure the statistical performance of the estimated PLS models with $K$ segments. The Akaike information criterion (AIC), the Bayesian information criterion (BIC), and the entropy statistic (EN), as recommended by Ringle et al. (2005), are used to measure the degree of separation. Ramaswamy, Desarbo, Reibstein, and Robinson (1993) also used EN to indicate the degree of separation in the estimated individual segment probabilities. The statistic bounded inclusively between zero and one is computed by Equation 4.7:

$$EN_K = \frac{1 - \left[ \sum_{i=1}^{N} \sum_{k=1}^{K} (-P_{ik} \ln P_{ik}) \right]}{(N \ln K)}, \quad (4.7)$$

where $EN_K$ is the entropy for the PLS model with $K$ bidder segments. Basically, a larger EN indicates a more crystallized or defined segmentation.

Although determining an appropriate value for $K$ is very important (Sarstedt et al., 2011), its actual determination is often more an art than a pure science. In general, low AIC, BIC, and CAIC, and high EN values indicate proper segmentation. However, these criteria are incommensurate. Low BIC values are usually only associated with one or two segment linear regression models. In addition to these criteria, other factors considered are the size of each segment and the number of significant variables. For example, if AIC, BIC, and CAIC are all lowest when $K = 2$ but there are also a large number of significant variables, it may be better to search for a better fit with $K > 2$. Currently, there are not any universally accepted criteria for FIMIX-PLS (Sarstedt et al., 2011).

## Response-Based Unit Segmentation PLS (REBUS-PLS)

Trinchera (2007) and Esposito Vinzi, Trinchrea, Squillacciotti, and Tenenhaus (2008) proposed another method for PLS-PM analysis to identify unobserved heterogeneity in the underlying population. Similar to FIMIX-PLS, response-based unit segmentation PLS (REBUS-PLS) assigns units (observations) to segments and estimates model parameters at the same time. Another similarity between FIMIX-PLS and this method is that both rely on a predefined theoretical model and do not require *a priori* knowledge of the differences between segments. Moreover, this integrated method practically eliminates an additional layer of potential errors stemming from the step of segmentation using estimated PLS scores or path coefficients.

Although FIMIX has been widely adopted to solve real-world business problems (e.g., Mancha, Leung, Clark, & Sun, 2014; Ringle et al. 2009; Sarstedt & Ringle, 2010), the method requires the input of the number of segments $K$ in the statistical computation (Mehmetoglu, 2011). Mancha et al. (2014) suggested an iteration through the finite state space of $K = 2, \ldots, k$ as a means to bypass this issue. Nevertheless, the utilization of this iterative process can be cumbersome, and the selection of the final FIMIX-PLS model (including the local models) may be ambiguous if the fit indices (such as goodness of fit) of models of different $K$ are relatively close. In light of this, Mehmetoglu (2011) and Esposito Vinzi et al. (2010) recommended the use of REBUS-PLS for dealing with heterogeneity in PLS-PM. The method, unlike FIMIX-PLS, analyzes the data and generates a dendrogram (tree graph) for segmentation of the customer population into subgroups. Based on the structure of the graph, the appropriate number of segments, $K$, can be identified, serving as the basis for REBUS-PLS' calculation of estimated path coefficients and assignment of units to segments. Their studies also claimed that REBUS-PLS provides an advantage through the use of the closeness measure (CM) in the computational procedure. According to Esposito Vinzi et al. (2010), the development of this new measure is based on the underlying beliefs that units belonging to the same latent class will have similar models and that if a unit is assigned to the correct latent class, its performance in the local model computed for that specific class will be better than the performance of the same unit considered as supplementary in the other local models.

As stated, the core of REBUS-PLS centers on CM between units and models in both measurement and structural models during the clustering. Through the examination of residuals related to the estimated measurement and structural models in the PLS construct, the procedure detects latent classes (if they exist) along with respective units sharing similar models. Its mathematical form is shown below in Equation 4.8:

$$CM_{nk} = \frac{\sum_{q=1}^{Q}\sum_{p=1}^{P_q}\left[\frac{e_{npqk}^2}{com(\hat{\xi}_{qk,x_{pq}})}\right] \times \sum_{j=1}^{J}\left[\frac{f_{njk}^2}{R^2(\hat{\xi}_j,\hat{\xi}_{q:\xi_q\to\xi_j})}\right]}{\sqrt{\frac{\sum_{n=1}^{N}\sum_{q=1}^{Q}\sum_{p=1}^{P_q}\left[\frac{e_{npqk}^2}{com(\hat{\xi}_{qk,x_{pq}})}\right]}{(N-t_k-1)} \quad \frac{\sum_{n=1}^{N}\sum_{j=1}^{J}\left[\frac{f_{njk}^2}{R^2(\hat{\xi}_j,\hat{\xi}_{q:\xi_q\to\xi_j})}\right]}{(N-t_k-1)}}} \tag{4.8}$$

where

*com* (•) is the communality index for the *p*th manifest variable of the *q*th block in the *k*th latent class;

$e_{npqk}$ is the measurement model residual for the *n*th unit in the *k*th latent class, corresponding to the *p*th manifest variable in the *q*th block—that is, the communality residuals;

$j_{njk}$ is the structural model residual for the *n*th unit in the *k*th latent class, corresponding to the *j*-endogeneous block;

*N* is the total number of units; and

$t_k$ is the number of extract components.

Essentially, CM is decided according to the structure of the goodness of fit (GoF) index, which is a geometric mean of average communality and the average $R^2$ values (Esposito Vinzi & Trinchera, 2008; Mehmetoglu, 2011). It should be noted that GoF index is the only measure for assessing the overall (global) fit of a PLS-PM model. Hence, a PLS model with a higher GoF index demonstrates a better overall fit on measurement and structural models (in an aggregate or global sense). Esposito Vinzi et al. (2010) suggested that a model with relative GoF index equal to or larger than 0.90 to be good. Readers can refer to the same article for a detailed technical explanation of the REBUS-PLS procedure.

It is important to note that integrated methods such as FIMIX-PLS and REBUS-PLS take into consideration of both inner (structural) and outer (measurement) models simultaneously during the estimation. On the other hand, clustering-based methods such as HC-PLS focus on only the latent variable in the inner model. Therefore, the segmentation by HC-PLS does not truly involve a global optimization of parameters capturing all measurements. Theoretically, FIMIX-PLS should yield a better fit than HC-PLS in terms of representing multiple segments within the PLS-PM model.

### Traditional Hierarchical Clustering

For illustrative purposes, this chapter contains an empirical analysis of online auction data using the three described methods, namely, HC-PLS, FIMIX-PLS, and REBUS-PLS. In order to provide a baseline for benchmarking, we also include the traditional hierarchical clustering (HC) in the segmentation example. In this conventional approach, indicator variables are first combined to form the latent variables in the measurement (outer) models. Then, HC is performed based on the observations' calculated and normalized latent variables. The underlying concept of the traditional HC method here parallels that in the HC-PLS method. The final selected model with $K$ segments indicates the unit assignment. It is imperative to point out that the traditional HC method is a naïve approach and does not take into consideration of the causality relationships embedded in the PLS-PM model. The segmentation simply treats the latent variables as independent inputs to the multivariate analysis. Because of its limitation in handling causal models, it is used as a raw basis in comparison of different methods.

## AN ILLUSTRATIVE EMPIRICAL ANALYSIS OF ELECTRONIC CUSTOMER DATA

As a demonstration, this section contains an empirical analysis of online auction data using the three methods described in the last section: hierarchical clustering of PLS scores (HC-PLS), finite mixtures partial least squares (FIMIX-PLS), and response-based unit segmentation partial least squares (REBUS-PLS). Hierarchical clustering (HC) on the normalized variables is first run to stablish a baseline. The methods selected exhibit different level of sophistication. As discussed, the first HC method does not take into consideration a causal model, while HC-PLS segments scores obtained on a PLS procedure, hence identifying segments accounting for each observation's path coefficients. FIMIX-PLS and REBUS-PLS, in addition, segment observations while optimizing the full model (i.e., both measurement and inner models).

To evaluate the four methods, we use a dataset consisting of 263 observations of online auction participants. For each participant, the dataset includes the level of enjoyment derived from participating in the auction, the level of uncertainty perceived from participating in the auction, the intention of using the same (or a similar) auction environment in the future, and the last posted bid. The first three variables were captured using a survey instrument (with four, three, and one items, respectively), answered right after the auction had finished. Although the model we evaluated has been built attending to a broad literature and sound theoretical principles, discussing them is out of the scope of this demonstration.

The theoretical model used to evaluate the segmentation methods based on an underlying model (i.e., HC-PLS, REBUS-PLS and FIMIS-PLS) is captured in Figure 4.2. The methods HC, HC-PLS, and REBUS-PLS were run using R (plspm package; Sanchez, Trinchera, & Russolillo, 2013). The FIMIX-PLS method was run using SmartPLS software (Ringle, Wende, & Becker, 2014). Post-hoc analyses of each segment were completed on R (plspm package), and 2,000 bootstrap samples used to estimate the significance of path coefficients.

We provide the GoF for each calculated model, yet the observed values are below the recommended fit threshold. The unadjusted GoF is sensitive to sample size and the distribution of the latent variables (i.e., violation of the multivariate normality) and should not be used as the sole metric to determine model fit, even less when segmenting a sample into smaller subgroups. Given the model comparison scenario under investigation, incremental (or relative) fit indices and those adjusted by sample size may be more suitable. Given the importance of comparing model fits and alternative models, we recommend the reader to explore and select a variety of fit metrics to analyze the resulting segments' models. The articles by Hu and Bentler (1995) and Hoyle and Panter (1995) offer a good overview on the topic and could be considered as the starting point when selecting appropriate and varied fit indices. Full evaluation of the segments, in addition to comparing the inner and outer models' path parameters and the model fit

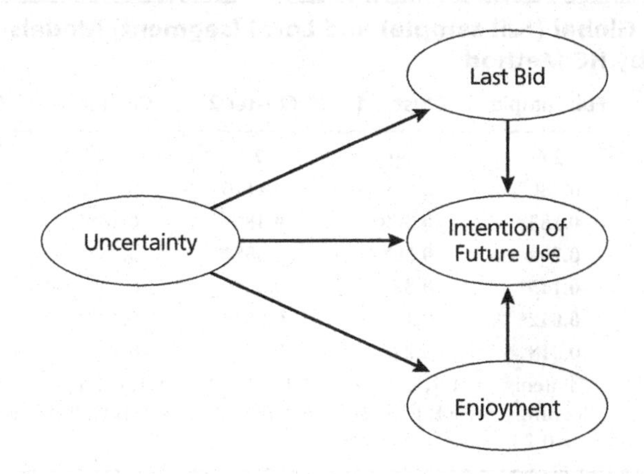


**Figure 4.2** Theoretical model.

indices, should study the distribution of observed variables for each identified segment.

## Hierarchical Clustering (HC)

To segment the observations using hierarchical clustering, the variables *enjoyment* and *uncertainty* were calculated as the addition of their items' scores. Then, all the variables were standardized. The application of the hierarchical clustering algorithm to the dataset resulted in the dendogram presented on Figure 4.3.

The HC method, exploratory in nature, resulted in the identification of four clusters. The PLS method was then applied to the observations on each segment to characterize them. Table 4.1 summarizes the segment sizes and the PLS path coefficients observed. It can be seen that the estimated inner paths of the PLS model of each segment are substantially different.

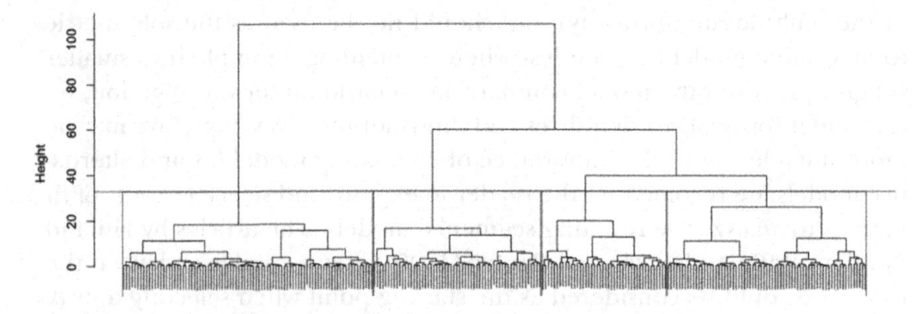


**Figure 4.3** Dendogram with four segments obtained using the HC method.

**TABLE 4.1 Global (full sample) and Local (segment) Models Estimated by HC Method**

| | Full sample | Cluster 1 | Cluster 2 | Cluster 3 | Cluster 4 |
|---|---|---|---|---|---|
| Size | 263 | 59 | 71 | 44 | 89 |
| GoF | 0.3867 | 0.3599 | 0.4176 | 0.3762 | 0.2614 |
| Unc->Enj | **–0.183*** | **0.369*** | **0.483*** | **–0.488*** | 0.185 |
| Unc->Bid | **0.292*** | **0.563*** | **0.398*** | 0.034 | 0.004 |
| Unc->IFU | **–0.145*** | **0.391*** | **0.338*** | –0.035 | 0.017 |
| Enj->IFU | **0.642*** | –0.140 | **0.280*** | **0.497*** | **0.509*** |
| Bid->IFU | –0.018 | –0.246 | 0.447 | –0.153 | 0.060 |
| Items retained on the outer model | All items loadings ≥ 0.7 | U1, U2, E1–E4, IFU, Bid | U1, U3, E1–E4, IFU, Bid | U1, U3, E1–E4, IFU, Bid | U2, U3, E1–E4, IFU, Bid |

* Significant path parameters with p-value ≤ 0.05

Those significant inner path coefficients with opposing signs are of particular interest, as the groups exhibit very different relationships between latent variables. As a consequence, the four identified segments displayed different customer profiles.

Simple interpretation of the results on Table 4.1 yields interesting findings. Clusters 1 and 2 exhibit similar significant relationships between the latent variables *uncertainty, enjoyment* and *intention of future use,* only different on the size of the parameters. For them, uncertainty increases enjoyment, bid amounts, and intention of future use, indicating the presence of a hedonic component and a preference for thrill. In contrast, cluster 3 has future use affected by enjoyment, which is reduced by greater uncertainty. In sum, the study of path coefficients can offer strategies for differential customer treatment.

## HC-PLS

This segmentation approach consists of first running the PLS-PM algorithm according to the theoretical model (Figure 4.1) and then applying the hierarchical clustering (HC) method to the latent variables scores obtained using PLS. Details of the approach have been described in Section 2.2. As shown in Figure 4.4, the dendogram generated by this method suggest that the examined customer base can be divided into two to four major segments (i.e., $K = 4$) and still end with segment sizes that can be analyzed. We chose four segments. This segmentation scheme is also displayed in that dendrogram. Each of the identified segments is described by a substantially different set of inner path causal relationships in the context of PLS, indicating dissimilar customer types (segments) within the heterogeneous population of online bidders (Table 4.2).

The results obtained with the HC-PLS method are quite dissimilar to those obtained with the HC method. The fact that the measurement model

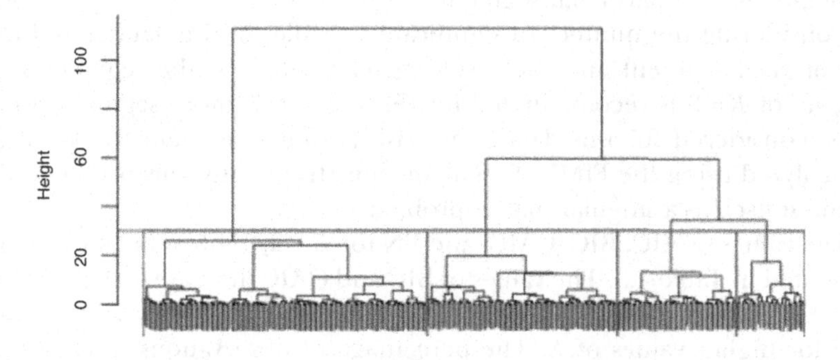


**Figure 4.4** Dendogram with four segments obtained using the HC-PLS method.

**TABLE 4.2 Global (Full Sample) and Local (Segment) Models Estimated by HC-PLS Method**

| | Full sample | Cluster 1 | Cluster 2 | Cluster 3 | Cluster 4 |
|---|---|---|---|---|---|
| Size | 263 | 76 | 113 | 27 | 47 |
| GoF | 0.3867 | 0.2827 | 0.2534 | 0.4630 | 0.2403 |
| Unc->Enj | **–0.183*** | **0.240*** | 0.387 | –0.183 | 0.205 |
| Unc->Bid | **0.292*** | **0.496*** | 0.164 | **0.439*** | –0.205 |
| Unc->IFU | **–0.145*** | 0.046 | –0.051 | 0.221 | **0.413*** |
| Enj->IFU | **0.642*** | 0.140 | **0.480*** | **0.650*** | –0.186 |
| Bid->IFU | –0.018 | –0.208 | 0.020 | –0.185 | 0.133 |
| Outer model—Items retained[a] | All items loadings ≥ 0.7 | U1, U2, E1-E4, IFU, Bid | U3, E1-E4, IFU, Bid | U1, U2, E1-E4, IFU, Bid | All items loadings ≥ 0.7 |

* Significant path parameters with p-value ≤ 0.05

[a] For some segments, lack of validity of their measurement model is a problem when using HC-PLS. For example, Cluster 3 has only one of its items for the latent variable Uncertainty showing the minimum required load.

is not taken into consideration by the segmentation process can generate problems of construct validity.

## FIMIX-PLS

As an illustration, the FIMIX-PLS procedure described earlier was used to process and analyze the auction dataset. Already explained in the method section, FIMIX-PLS procedure requires an examination of the results for different values of *K* in an iterative manner. The ideal number of segments *K* for classification of bidders was identified using criteria measuring the degree of separation—that is, AIC, BIC, CAIC, and EN. Subsequently, a bidder is assigned to the segment with the highest estimated probability of belonging to that particular segment.

Considering the number of significant variables in the dataset and the size of each segment and the resulting segment sizes, the segmentation scheme of $K = 3$ is recommended for this dataset. If more segments were to be considered for this dataset, we risk having a segment too small to be analyzed using the FIMIX-PLS algorithm (frequently, this problem will manifest itself as a singular matrix problem).

The values of AIC, BIC, CAIC, and EN for $K = 2, 3, 4$ for the auction are presented in Table 4.3. The values of BIC and CAIC increase with the value of *K*. EN is the highest (0.704) when $K = 2$, but fluctuates between 0.55 and 0.56 for higher values of *K*. The percentage of observations in each segment ranges from 15.9% to 54.8%. Estimated path coefficients as well as the

**TABLE 4.3 Segmentation of Customer Data Using FIMIX-PLS**

| *K* | 2 | 3 | 4 |
|---|---|---|---|
| AIC | 2,009.63 | 1,986,20 | 1,983.88 |
| BIC | 2,070.36 | 2,079.08 | 2,108.91 |
| CAIC | 2,087.36 | 2,105.08 | 2,143.91 |
| EN | 0.704 | 0.552 | 0.560 |

**TABLE 4.4 Global (Whole Sample) and Local (Segment) Models Estimated by FIMIX-PLS Method**

| | Full sample | Segment 1 | Segment 2 | Segment 3 |
|---|---|---|---|---|
| Size | 263 | 53 | 161 | 49 |
| GoF | 0.3867 | 0.4929 | 0.3671 | 0.6284 |
| Unc->Enj | **–0.183*** | **0.376*** | **–0.361*** | **–0.588*** |
| Unc->Bid | **0.292*** | 0.648 | **0.310*** | **0.476*** |
| Unc->IFU | **–0.145*** | **0.294*** | **–0.391*** | –0.164 |
| Enj->IFU | **0.642*** | **0.724*** | **0.332*** | **0.944** |
| Bid->IFU | –0.018 | –0.093 | 0.232 | **–0.121*** |
| Outer model—Items retained | All items loadings ≥ 0.7 | All items loadings ≥ 0.7 | All items loadings ≥ 0.7 | All items loadings ≥ 0.7 |

* Significant path parameters with p-value ≤ 0.05

sizes and goodness of fit (GoF) indices for the global and local models are tabulated in Table 4.4. A global model captures the entire sample dataset, whereas the local model represents each segment identified by the FIMIX segmentation method. It is apparent that the construct of the inner path models for the three segments as well as the global model consisting of all customers are statistically different from each other.

The interpretation of path coefficients for the different segments is similar to the one offered for the HC method. It can be observed that the estimated inner path structure in the global PLS-PM model is statistically different from those of the three local models. Once again, we observe different signs in the path coefficients across segments, indicating that very different processes are at play.

## REBUS-PLS

The customer dataset used in the above methods was also analyzed using REBUS-PLS. The procedure created the dendrogram depicted in Figure 4.5. Analysis of separation of the outer and inner residuals in the model indicates

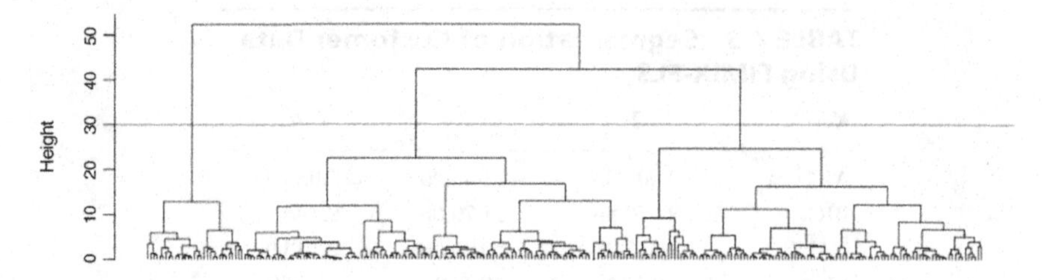


**Figure 4.5** Dendogram with four segments obtained using the REBUS-PLS method.

**TABLE 4.5 Global (Full Sample) and Local (Segment) Models Estimated by REBUS-PLS Method**

| | Full sample | Segment 1 | Segment 2 | Segment 3 |
|---|---|---|---|---|
| Size | 263 | 75 | 94 | 94 |
| GoF | 0.3867 | 0.2397 | 0.2330 | 0.2599 |
| Unc->Enj | **–0.183*** | 0.136 | –0.080 | –0.186 |
| Unc->Bid | **0.292*** | **0.286*** | **0.282*** | **0.357*** |
| Unc->IFU | **–0.145*** | **–0.290*** | –0.099 | –0.177 |
| Enj->IFU | **0.642*** | **0.354*** | **0.342*** | **0.312*** |
| Bid->IFU | –0.018 | 0.086 | –0.104 | –0.022 |
| Outer model—Items retained | All items loadings ≥ 0.7 | All items loadings ≥ 0.7 | All items loadings ≥ 0.7 | U1, U2, E1, E2, IFU, Bid |

* Significant path parameters with p-value ≤ 0.05

that the customer base is most appropriately divided into three segments. REBUS then classified the units into segments according to the specified $K = 3$ scheme. Once the observations were assigned to one of the three clusters, PLS was then run again on each one of the three segments. Results of the global (entire sample data) and local (segment) models are provided in Table 4.5.

It can be observed that the estimated inner path structure in the global PLS-PM model is statistically different from those of the three local models. Furthermore, a comparison of Table 4.5 (for REBUS-PLS) with Tables 4.2 (HC-PLS) and 4.4 (FIMIX-PLS) indicates that different methods may lead to potentially conflicting conclusions (such as different segmentation schemes and assignments of units to segments) and interpretations of statistical modeling results.

## CONCLUSION

Customer analytics often involves the examination of heterogeneous segments within a population. This unobserved heterogeneity in customer data may cause errors in establishing the causal relationships in a PLS model and potentially lead to misinterpretation of results. In this chapter, we review three methods for PLS modeling and segmentation and explain how to identify customer segments with dissimilar consumer characteristics and behaviors. For illustrative purposes, our chapter also contains an empirical analysis of online auction customer data using these three methods, along with a hierarchical clustering benchmark. Results indicate that different methods may result in dissimilar segmentation schemes and varied conclusions, attending to a theoretical model of outer (measurement) and inner (structural) relationships. Among the methods explored, FIMIX-PLS offers the segmentation algorithm offering best-fitting models, offering clear insight into the nature of the identified segments, and it does not negatively affect the measurement model. Due to the possibility of conflicting guidance by different methods, we suggest that further research is needed to develop a unified metric for comparison of the efficacy of these methods and to enhance their computational efficiency. More tests utilizing large datasets collected from real-world environments are also needed to establish the relative strength and weakness of these methods.

## REFERENCES

Esposito Vinzi, V., & Trinchera, L. (2008). Latent class detection in component-based structural equation modeling. *Proceedings of the XLIV SIS Scientific Meeting, XLIV Scientific Meeting of the Italian Statistical Society (SIS)* (pp. 147–154). Invited paper session: Padova, Italy.

Esposito Vinzi, V., Trinchera, L., Squillacciotti, S., & Tenenhaus, M. (2008). REBUS-PLS: A response based procedure for detecting unit segments in PLS path modeling. *Applied Stochastic Models in Business and Industry, 24*, 439–458.

Esposito Vinzi, V., Trinchera, L., & Silvano, A. (2010). PLS path modeling: From foundations to recent developments and open issues for model assessment and improvement. In V. Esposito Vinzi, W. W. Chin, J. Henseler, & H. Wong (Eds.), *Handbook of partial least squares* (pp. 47–82). Berlin, Germany: Springer-Verlag.

Hahn, C., Johnson, M.D., Herrmann, A., & Huber, F. (2002). Capturing customer heterogeneity using a finite mixture PLS approach. *Schmalenbach Business Review, 54*, 243–269.

Hair, J. F., Anderson, R. E., Tatham, R. L., & Black, W. C. (1998). *Multivariate data analysis* (5th ed.). Upper Saddle River, NJ: Prentice Hall.

Hoyle, R. H., & Panter, A. T. (1995). Writing about structural equation models. In R. H. Hoyle (Ed.), *Structural equation modeling: Comments, issues, and applications* (pp. 158–176). Thousand Oaks, CA: Sage.

Hu, L. T., & Bentler, P. M. (1995). Evaluating model fit. In R. H. Hoyle (Ed.), *Structural equation modeling: Concepts, issues, and applications* (pp. 76–99). Thousand Oaks, CA: Sage.

Hutchinson, J. W., Kamakura, W. A., & Lynch, J. G. (2000). Unobserved heterogeneity as an alternative explanation for "reversal" effects in behavioral research. *Journal of Consumer Research, 27,* 324–344.

Kuo, R. J., Ho, L. M., & Hu, C. M. (2002). Integration of self-organizing feature map and K-means algorithm for market segmentation. *Computers and Operations Research, 29*(11), 1475–1493.

Lyk-Jensen, S. V., & Chanel, O. (2007). Retailers and consumers in sequential auctions of collectibles. *Canadian Journal of Economics 40,* 278–295.

Mancha, R., Leung, M. T., Clark, J., & Sun, M. (2014). Finite mixture partial least squares for segmentation and behavioral characterization of auction bidders. *Decision Support Systems, 57,* 200–211.

Mehmetoglu, M. (2011). Model-based post hoc segmentation (with REBUS-PLS) for capturing heterogeneous consumer behavior. *Journal of Targeting, Measurement and Analysis for Marketing, 19,* 165–172.

Ramaswamy, V. Desarbo, W. S., Reibstein, D. J., & Robinson, W. T. (1993). An empirical pooling approach for estimating mix elasticies with PIMS data. Marketing Science, 12, 103–124.

Ringle, C. M., Sarstedt M., & Mooi, E. A. (2009). Response-based segmentation using finite mixture partial least squares: Theoretical foundations and an application to American customer satisfaction index data. *Annals of Information Systems, 8,* 19–49.

Ringle, C. M., Wende, S., & Becker, J. M. (2014). *SmartPLS 3.* Hamburg, Germany: SmartPLS. Retrieved from http://www.smartpls.com

Ringle, C. M., Wende, S., & Will, A. (2005). Customer segmentation with FIMIX-PLS. In T. Aluja, J. Casanovas, V. Esposito Vinzi, A. Morineau, & M. Tenenhaus (Eds.), *Proceedings of PLS-05 International Symposium, Paris,* pp. 507–514.

Sanchez, G., Trinchera, L., & Russolillo, G. (2013). *Package PLSPM: Tools for Partial Least Squares Path Modeling. R package version 0.4-1.* Retrieved from: http://CRAN.R-project.org/package=plspm

Sarstedt, M. (2008). A review of recent approaches for capturing heterogeneity in partial least squares path modeling. *Journal of Modelling in Management, 3,* 140–161.

Sarstedt, M., Becker, J. M., Ringle, C. M., & Schwaiger, M. (2011). Uncovering and treating unobserved heterogeneity with FIMIX-PLS: Which model selection criterion provides an appropriate number of segments? *Schmalenbach Business Review, 63,* 36–62.

Sarstedt, M., & Ringle, C. M. (2010). Treating unobserved heterogeneity in PLS path modeling: A comparison of FIMIX-PLS with different data analysis strategies. *Journal of Applied Statistics, 37,* 1299–1318.

Sarstedt, M., Schwaiger, M., & Ringle, C. M. (2009). Do we fully understand the critical success factors of customer satisfaction with industrial goods? Extending Festge and Schwaiger's model to account for unobserved heterogeneity. *Journal of Business Marketing Management, 3,* 185–206.

Tenenhaus, M., Esposito Vinzi, V., Chatelin, Y., & Lauro, C. (2005). PLS path modeling. *Computational Statistics and Data Analysis, 48,* 159–205.

Trinchera, L. (2007). *Unobserved heterogeneity in structural equation models: a new approach in latent class detection in PLS path modeling.* Unpublished PhD thesis, DMS, University of Naples, Italy.

Wold, H. (1975a). PLS path models with latent variables: The nipals approach. In H. M. Blalock, A. Aganbegian, F. M. Borodkin, R. Boudon, & V. Cappecchi (Eds.), *Quantitative sociology: International perspectives on mathematical and statistical modeling* (pp. 47–82). New York, NY: Academic Press.

Wold, H. (1975b, August). *Modelling in complex situations with soft information.* Paper presented at the Third World Congress of Econometric Society, Toronto, Canada.

# SECTION II

BUSINESS APPLICATIONS

CHAPTER 5

# TEACHING A DATA MINING COURSE IN A BUSINESS SCHOOL

**Ronald K. Klimberg**
***Saint Joseph's University***

## ABSTRACT

New business intelligence/business analytics (BI/BA) majors and minors at the undergraduate and MBA levels and new master's programs are appearing in business schools. Most of these programs include a data mining course. The emphases of these courses and programs vary greatly. A major source of this variation is due to institution's definition of business intelligence (BI) and business analytics (BA). Initially in this chapter, we develop an integrated framework for business intelligence and business analytics. Based on our framework and recent survey and interviews of BI/BA programs, we identify what we believe are the critical fundamental components that should exist for successful BI/BA programs and in particular in a data mining course offered in a business school.

*Contemporary Perspectives in Data Mining, Volume 2*, pages 79–95

## INTRODUCTION

The business world is embracing the business intelligence/business analytics (BI/BA) revolution. The academic world is reacting by offering undergraduate majors and minors, Masters of Science degrees, certificates and concentrations within MBA programs. The emphasis of these courses and programs vary greatly. A major source of this variation is due to institution's definition of BI and BA. In the next section of the chapter, we will discuss these various point of views of business intelligence (BI) and business analytics (BA) and develop an integrated framework. Subsequently, we identify what fundamental analytic skills all business students should have and further identify in general the key analytic skills students majoring or minoring in BI/BA should have. Additionally, we focus on the analytic skills students should obtain from a data mining course.

## WHAT ARE BI AND BA?

Howard Dresner, considered by many as the father of business intelligence, used the term business intelligence (BI) to describe the area of "concepts and methods to improve business decision making by using fact-based support systems" in 1989 (Evans, 2010, para. 9).[1] Nonetheless, it was not until the late 1990s that the term *business intelligence* fell into widespread use. At that time, many companies motivated by the fear of Y2K issues upgraded their databases or made significant investments in new data warehouses and ERP systems. With additional data more readily available, a renewed interest grew in developing decision support systems to retrieve and report the information. Meanwhile in academia, a few BI courses started to appear, mostly in information systems (IS) departments. These courses were typically focused on decision support systems, query, reporting and dashboards (OLAP tools), and information systems, from databases, data warehouses, data marts and ERP systems.

Since about the start of the 21st century, the BI field as well as its definition has expanded, mainly fueled by the exponentially increasing capabilities of technology. As futurist Ray Kurzweil stated:

> The computer in your cell phone today is a million times cheaper and a thousand times more powerful and about a hundred thousand times smaller (than the one computer at MIT in 1965) and so that's a billion-fold increase in capability per dollar or per euro that we've actually seen in the last 40 years. (Lomas, 2008, para. 4)

The equivalent of Moore's law of computing has occurred in data storage in terms of speed and costs. With this new and ever improving technology,

> most organizations (and even small organizations) are collecting an enormous amount of data.... The IT budget for most organizations is a significant percentage of the organization's overall budget and is growing.... No matter if the size of the organization is large or small, only a limited number of organizations, (yet growing in number), are using their data extensively. (Klimberg & McCullough, 2013b, p. 2)

Significantly expanded analytical capabilities are now available with new and enhanced software. With these changes, the area of BI and its definition have evolved so as to now include not only the gathering, storing, reporting and providing data, but also including analyzing the data using statistical and quantitative methods to assist users to make better business decisions.

One of the first areas of significant expansion of BI was the use of advanced statistical methods such as data mining. Data mining techniques are able to sift through large amounts of data to find unsuspected relationships. Industry has applied these advanced statistical techniques to target marketing, market segmentation, market basket analysis, risk management, forecasting, customer retention, fraud detection, and so on. Academic institutions in turn reacted by starting to offer new data mining courses as required or elective courses as part of their BI programs.

A second area of BI's expansion has been the application of quantitative techniques, such as optimization and simulation. This second area of BI's expansion has revived an old term, business analytics (BA).[2] Davenport and Harris in their 2007 book, *Competing on Analytics*, describe business analytics as "the extensive use of data, statistical and quantitative analysis, explanatory and predictive models, and fact-based management to drive decisions and actions" (Davenport & Harris, 2007, p. 7). The Institute for Operations Research and the Management Sciences (INFORMS), the largest professional society in the world for professionals in the field of operations research (OR), management science (MS), and analytics, has recommended a general definition of analytics as "the scientific process of transforming data into insight for making better decision" (INFORMS, 2013a, para. 1). The INFORMS analytics section further decomposes their broad analytics definition into three categories:

**Descriptive analytics**

- Prepares and analyzes *historical* data
- Identifies patterns from samples for reporting of trends

**Predictive analytics**

- Predicts *future* probabilities and trends
- Finds relationships in data that may not be readily apparent with descriptive analysis

**Prescriptive analytics**

- Evaluates and determines *new* ways to operate
- Targets business objectives
- Balances all constraints (INFORMS 2013b).

The graphical view in Figure 5.1 illustrates the range of analytics and the type of analytics as the level of intelligence varies (Davenport & Harris, 2007). The area below the horizontal line is where BI originated in accessing and reporting data. In context of the INFORMS analytics definition, this area would be descriptive analytics. Descriptive analytics consists of reporting, visualizing, and slicing and dicing the data. The area above the horizontal line, which now includes BI's expanded definition, represents the tools of advanced predictive and prescriptive analytics. Predictive analytics employs advanced statistical techniques such as data mining to uncover relationships not readily apparent with descriptive analytics and predicts the future, while prescriptive analytics utilizes optimization, simulation, and other quantitative tools to improve the organizational efficiency. Organizations and members of organizations such as INFORMS are more focused on the advanced predictive and prescriptive analytics.

The field of BI and its definition has evolved from its early definition of descriptive analytics to now include predictive and prescriptive analytics. Meanwhile, in recent years, the term *business analytics* (BA) has come to be used to describe BI's core reporting and descriptive analytics along with advanced statistical methods and quantitative techniques—that is, advanced

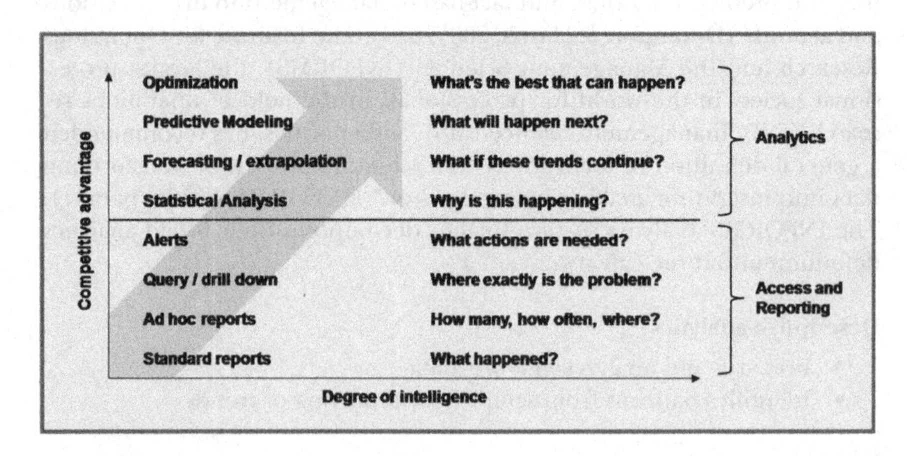


**Figure 5.1** Variation of BI/BA through the degrees of intelligence of the methodologies employed as it directly affects competitive advantage. *Source:* Davenport, T. H. & Harris, J. G. (2007). *Competing on Analytics.* Boston, MA: Harvard Business School Press.

predictive and prescriptive analytics. These definitions and viewpoints leave us with several perplexing questions:

- Exactly what are BI and BA?
- How are they related?
- How do they differ?

BI and BA seem somewhat similar, yet some people view BI as a subset of BA while other people view BA as a subset of BI. The point of view one has seems to depend upon mostly on one's training. For example, if individuals are more from the information systems (or in academia from the MIS, IS, or IT-like departments), they would view BA as a subset of BI. On the other hand, if the people are more from the quantitative, statistical world (or in academia from the DSS, MS, OR, statistics, or QM-like departments), they view BI as a subset of BA. Since BA is currently the more dominant term, in this chapter we will view BI as a subset of BA.

Analytics has several stages over which it can vary: either by analytic category, using INFORMS definition, from descriptive to predictive to prescriptive, or by level of intelligence/level of competitive advantage as viewed by Figure 5.1. Regardless, a discipline-based framework of analytics that encapsulates these viewpoints includes the three core subject areas: information systems and technology, statistics, and quantitative methods, as shown in Figure 5.2 (adapted from Klimberg & Miori, 2012). With our view of the term BA, we define BA as the combination of information systems/technology

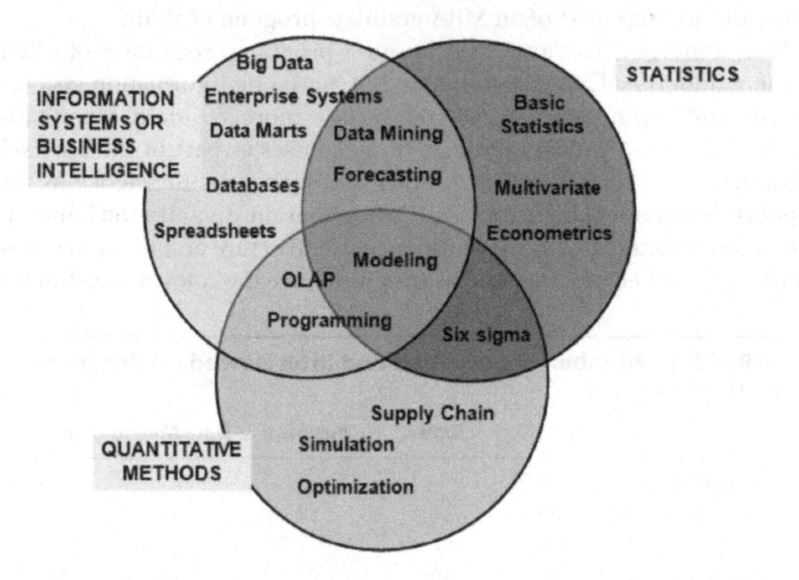


**Figure 5.2** A discipline-based approach to BA.

with statistics and quantitative methods. The area above the horizontal line in Figure 5.1 in context of our Venn diagram of BA is advanced statistical or quantitative techniques applied to traditional BI systems, extending and enhancing traditional BI capabilities.

Irrespective of your point of view, the bottom line is that BA and BI both apply "data, technology, and analytics to gain insight and knowledge that enables decisions about processes, services, and products that yield positive economic outcomes" (Herschel, 2010, para. 18).[3] McKinsey's research "suggests that we are on the cusp of a tremendous wave of innovation, productivity, and growth, as well as new modes of competition and value capture—all driven by big data as consumers, companies, and economic sector exploit its potential" (Manyika et al., 2011, p. 12). An increasing number of organizations and academic institutions are engaging in using BI/BA and are moving up the BI/BA maturity path.

## FUNDAMENTAL BA SKILLS FOR ALL BUSINESS STUDENTS

We recently surveyed and interviewed 32 institutions to obtain a sense of the current landscape of analytics programs (Gorman & Klimberg, forthcoming). As shown in Table 5.1, a total of 43 programs were identified. Most of programs were offered in the business school: 38 out of the 43. All nine undergraduate programs are majors in BA or BI and are part of a larger undergraduate business education. Similarly, the 12 concentrations or certificates were part of an MBA graduate program of study.

Most undergraduate and MBA business programs, regardless of whether the school is AACSB accredited or not, have one information systems/ information technology course and two or more required quantitative methods courses. AACSB identifies these courses as part of the core skill learning areas: "statistical data analysis and management science as they support decision-making processes throughout an organization" and "information technologies as they influence the structure and processes of organizations and economies, and as they influence the roles and techniques

**TABLE 5.1 Number of Surveyed and Interviewed BA Programs by Degree**

| | Total | Business | Data Mining Course |
|---|---|---|---|
| Undergraduate | 9 | 9 | 7 |
| Concentration or Certificate | 13 | 12 | 9 |
| Masters | 21 | 17 | 16 |
| Total | 43 | 38 | 32 |

of management" (AACSB, 2005, para. 11). These courses provide a basic introduction to the BA area. Further, several additional core required courses provide a fundamental understanding of business—courses such as accounting, marketing, finance, and management.

One of the major objectives of most business programs, which is not discipline-based but cuts across several courses, is the improvement of communication skills—written and oral. This skill is a critical element for the success of BA implementation. You could have some of the best analytics model/results, but they are doomed to failure and will never be used if they are not communicated properly to the decision makers that would use the model/results. The results must be written or presented to decision makers in a nontechnical form, using appropriate jargon and demonstrating an understanding of the problem situation. Good communication skills and a basic fundamental understanding of the business environment are essential components to maximize the likelihood the results from any analytics project being implemented. The need for these "soft skills" and the ability to conduct a nontechnical discussion and understanding of the problem situation is even more imperative today.

In addition, many academic institutions have begun to offer entirely new MS programs that specialize in BA. The 21 Masters of Science programs we surveyed are more novel in their structure and focus. Most of these MS programs appear to be offered in a business school; from our survey, 17 out the 21 MS programs were housed in a business school. Discipline-wise, it appears most of these programs focused on only one or two of the three major BA disciplines (IT, statistics, or quantitative methods). (An early survey found similar results—Klimberg & McCullough, 2013a.) We believe a BA program should not be envisioned in isolation of only one or two disciplines. Regardless of the focus of the program, course coverage of a BA program should include a foundational understanding of all three disciplines. As a result, the students would attain a better understanding of the integrative nature of BA skills. Additionally, most of these MS programs focus very little or not at all on the foundational skills of business and communication skills that the undergraduate and MBA programs do provide. These Master's programs in BA should also provide these foundation courses.

In summary, all general undergraduate and MBA business students, as well as all majors and minors in BA, and MS in BA students, should be provided the fundamental competency in:

- understanding business
- communication skills
- the introduction to the three discipline areas of BA

### Analytics Across the Curriculum

Wixom, Ariyachandra, Goul, Gray, Kulkarni, and Phillips-Wren (2011) stated in their article that the current state of BI in academia "may be behind the curve in delivering effective Business Intelligence programs and course offerings to students" (Wixom et al., 2011, p. 1). The same situation is true with BA—the current state of BA in academia lags behind industry. As we have observed from our survey many new majors, minors, and concentrations in BA and MS in BA programs have appeared. However, with some of these programs (not the MS in BA programs), we witnessed that this spike of interest has resulted with some institutions minimally rebranding themselves. These institutions "simply change the names of their courses and programs to now include the words business analytics or business intelligence in their names" (Klimberg & McCullough, 2013a, p. 57). This is quite similar to what we saw with the word "green" in many environmental academic programs and courses a few years ago. If this is what is going to be the major outcome of this BA/BI resurgence, we will lose an enormous opportunity.

In order for the ultimate success for business schools to be attained—to decisively achieve the basic functional knowledge and application of analytics—analytics must spread beyond the walls of the BA department and promulgate throughout the entire curriculum. Analytics should be provided across the entire curriculum. In order for this to take place, a significant paradigm shift must occur. This shift would be a significant culture change to most institutions. Analytics, at least at the descriptive analytics level, should be promoted and utilized in all the business discipline areas. In order for this change to occur, many professors will have to update/retool their skill set to utilize these analytic tools and techniques as part of their classes. The end result of this paradigm shift will be making analytical skills an essential set of fundamental capabilities in any business degree. This shift is much like what is happening in industry, and the demand is there. McKinsey's 2011 Big Data report (Manyika et al. 2011) stated that by 2018, the United States alone will face a shortage of 1.5 million managers and analysts to analyze big data and make decisions based on their findings.

## FUNDAMENTAL BA SKILLS FOR ALL BA STUDENTS

BA programs, regardless of their discipline(s) focus do provide the students with sufficient introduction to the BA tools, techniques, and technology. The reason for the emphasis on the tools, techniques, and technology is to learn the methodology/technology to developing models/tools to solve problems. Nonetheless, we do not spend enough time on the critical skill of problem solving or modeling, which is the process of structuring and analyzing problems to develop some quantitative abstraction of the problem that will lead to a rational course of action. The skills needed to analyze a problem situation and building a model are fundamental skills of scientists and engineers, as well as other disciplines, such as business managers, educators, and healthcare providers. In the classroom, the mathematics of the tools and techniques is emphasized, and not enough time is spent on the skills necessary to analyze a problem situation. "The teaching of models is not equivalent to the teaching of modeling" (Morris, 1967, p. B708). We need to teach BA students the craft of modeling. We have these problem-solving/modeling skills—the "Holy Grail" of skills and the most difficult to accomplish (Klimberg& McCullough 2013a). If students can master these problem-solving skills, they can solve any problem using any technology, tool, or technique.

A primary reason for our students' poor problem-solving skills is that students do not know where and when to use the tools and techniques we teach them, and more so, how to go about solving a real problem (Grossman, 2002; Powell, 2001). We may cover certain chapters that encompass a particular technique and then give them homework or test the students on that technique. And our students do fairly well. However, at the end of the semester, during a comprehensive final exam, most students have difficulty deciding what technique to use where and when. As an analogy, we teach our students how to use the hammer, drill, saw, and so on. But, they don't know anything about how to be a carpenter—given their tools and a stack of lumber and other materials, they have no idea where to start to build a house. The problem solver's ability to develop an effective representation and use a problem-solving tool also is a function of his or her level of expertise. Similarly, the subjects may be unable to appreciate the full value of hints about move operators because of unfamiliarity with the type of problem they are asked to solve. We teach our students the methodologies to apply the tools and techniques, yet we do not teach them the modeling strategies necessary to take them from novices to experts. We must teach our students the craft of modeling skills so that they can become master carpenters.

An important difference between experts and novices is that the experts tend to model the problem and its behaviors and work from that model, whereas novices need to work from familiar formulas and available variables (Larkin, McDermott, Simon, & Simon, 1980). Thus, a novice has difficulty getting and interpreting a "bird's-eye view." We extended the theory of cognitive fit (Vessey, 1991) to a three-way relationship among the decision maker's skill level (with both the technique and application area), the problem-solving task, the problem-solving tools, and problem representation, as shown in Figure 5.3 (Klimberg & Cohen, 1999). The extent to which this match occurs will increase the efficiency of the problem-solving process.

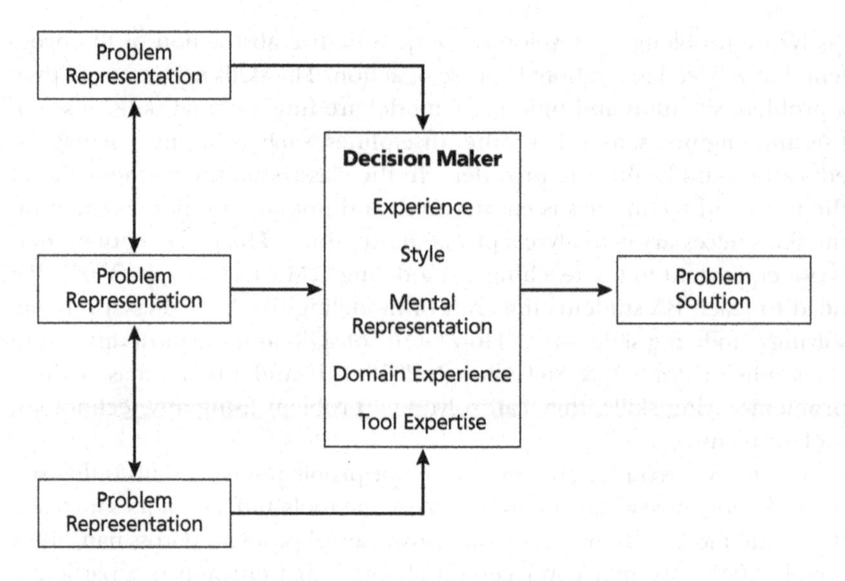


**Figure 5.3** Extended cognitive fit model to problem-solving.

The application of heuristics is one of the fundamental problem-solving strategies of experts (Schön, 1983). In an art of modeling course, Powell takes an engineering approach to practical problem-solving (Powell, 1995a, 1995b). The students develop insights and the skills and apply them to analyzing practical problem situations. Powell's art of modeling pedagogy follows the problem-based learning approach of the cognitive apprentice model (Collins, Brown, & Newman, 1989). Problem-based learning is an educational approach in which students acquire conceptual knowledge and practical skills to solve problems and improve comprehension. Students take on active learning strategies and develop critical thinking skills and are challenged to "learn to learn." Problem-based learning has thrived mainly in medical and engineering schools and has slowly been embraced by other schools. The cognitive apprentice model, developed by Collins, Brown, and Newman, has the students learning through coaching from the master. A significant component of the cognitive apprentice model is "scaffolding." Scaffolding is breaking down the tasks into manageable tasks and providing learning aids, such as heuristics, to facilitate and direct the students in achieving the learning goals of the task. The effectiveness of this model depends to a large degree on the scaffolding provided by the instructor.

Willemain conducted experiments eliciting verbal protocols from experts during the initial stages of model formulation (Willemain, 1994, 1995). "This research established the idea that experts use model structuring as a backbone on which to build, while making frequent side trips to consider issues such as the client's needs, implementation, data availability, and so on" (Powell & Willemain, 2003). Powell and Willemain (2003) proposed to conduct similar protocol experiments with novice MBA students. This research would improve "our understanding of the modeling process of novices" (Powell & Willemain, 2003).

The CONDOR Report (1988) called for the development of a "modeling science" to teaching modeling, especially teaching modeling to novice modelers. Powell's art of modeling approach emphasizes the teaching of the craft of modeling and improves the match between the decision maker (student) and the problem-solving task. The problem-solving tool that Powell expects his students, who are MBA students, to use to "solve" and represent the problem is a spreadsheet. In general, most MBA students are more familiar with "real-world" applications in spreadsheets and are more advanced novice modelers than undergraduate students. For most novice modelers, the spreadsheet is not the best problem-solving tool/problem representation. Although modelers, even novice modelers, can now easily develop models within a spreadsheet, the spreadsheet, especially beyond basic functionality, still can present a significant barrier to many novice student modelers. The teaching and development of modeling heuristics skills should improve the problem-solving process.

## Data Mining Course

One area of initial expansion of BI, as we discussed earlier, was to offer an advanced statistics course usually about data mining. Based on our survey, this statement is reaffirmed, as we observed that most of the programs did include a data mining course (Table 5.1). For most MS programs, the data mining course was a required course, while in most MBA programs the data mining course was one of the elective courses in the BI/BA concentration.

The course content of these data mining courses focuses on three main topic areas: data preparation and visualization, multivariate statistical techniques, and data mining techniques. The number of techniques covered in the course ranged from 8 to 15, and included:

- Data preparation and visualization: exploration, cleaning, preparation and visualization
- Multivariate statistical techniques: cluster analysis, principal component analysis, discriminant analysis, multiple regression, logistic regression[4]
- Data mining techniques: decision trees, neural networks, market basket analysis, k-means, naïve Bayes, text mining, and model comparison.

Typically, the only prerequisite for students to take the data mining class is an introductory-level business statistics class. So a student jumps from univariate/bivariate statistical analysis to statistical data mining techniques that include numerous variables and records. These statistical data mining techniques implicitly require the student to understand the fundamental principles of multivariate statistical analysis, and more so, to understand the process of a statistical study, which is directly related to the skill of problem-solving/modeling. Further, too many techniques are being covered; on average, the data mining course covers at least one technique a week. Together these factors produce a situation in which many students become lost and further heighten their view that the course is just another math course.

Generally, two approaches have been taken to try to improve student problem-solving/modeling skills in the data mining course. Recently, a novel approach that has been tried has been the flip classroom. Flip teaching is also called *reverse teaching* or *backwards classroom.* Instead of using the classroom time to lecture, students learn new content online by watching videos prior to class. The classroom time is used to go over assigned problems, which in a regular classroom would be the homework problems. Using the classroom time to go through the problems allows the students to actively learn and to receive more personalized guidance. Another more traditional approach to improve problem-solving/modeling skills has been to include end-of-semester (or semester-long) group projects to analyze a dataset using several of the techniques covered in the course. With so many techniques covered in the course, the dataset project customarily ends up being viewed as a statistics project. This is not necessarily a bad outcome, as the dataset project does take a major step to improving their problem-solving/modeling skills. Even so, a gap still remains between the techniques, problem-solving/modeling skills, and real-world decision-making situations. An approach to improving/solidifying their problem-solving skills and achieve a business application/modeling situation is to provide them with opportunities for experiential learning.

## Experiential Learning Experiences

Experiential learning (EL) is a process of learning by doing. The members of the Association for Business Simulation and Experiential Learning (ABSEL), an organization for promoting innovative and effective teaching methods, "have used the following quote, attributed to Confucius, to express their conviction that experiential learning is effective: I hear and I forget; I see and I remember; I do and I understand" (Gentry, 1990, p. 1). Students learn from interactive and applied experiences, as opposed to more traditional teaching methods of rote or didactic learning. "The focus of EL is placed on the process of learning and not the product of learning" (UC Davis, 2011, para. 6). Innovative EL courses have been successfully designed and implemented despite the challenges of integrating the requirements of both academia and industry. Stimulating EL experiences will significantly mature the student's problem-solving/modeling skills. An approach that meets the criteria of EL is a live case—that is, bringing a real-world problem situation into the classroom (Gentry, 1990).

An example of EL experience using analytics is when we used real-world data for medical and pharmacy claims and combined a pharmaceutical marketing and a BA class (Sillup, Klimberg, & McSweeney, 2010). In a collaborative process to mimic how the pharmaceutical industry determines the potential of new drug, integrated student teams worked together and taught each other their respective knowledge areas, such as marketing or statistics. This collaborative learning experience worked well. The students experienced an increase understanding of another discipline, but more so, they developed a deeper understanding of their own discipline. Further, they also learned to work constructively on teams consisting of members with different backgrounds, not unlike situations they will encounter in their professional careers within the corporate world, and to behave more like self-directed adult learners (Knowles, 1990).

We will start in the spring of 2014 to offer a continuing innovative EL experience using analytics. As part of our undergraduate data mining course at Saint Joseph's University (SJU), required by all our BA majors and minors, the students will participate in the SJU/SAP/POI Analytics Cup. The students will have an opportunity to use SAP's Trade Promotion Optimization (TPO) software package.

Trade promotion management is the process employed by the consumer product goods (CPG) industry to plan, budget, present, and execute incentive programs between manufacturers and retailers. TPO utilizes advanced analytics, such as optimization, simulation, and forecasting, to improve the TPM process by reducing forecasting error. For example, if we take Safeway, one of the largest retailers in the U.S., in 2011 they had $44,206,500,000 in net sales. If we assume a 2% ROI on TPO, that would be about $884,130,000. Since a significant portion of sales does not come from food sales, let's assume it takes 50% of the profit for a change of $442,605,000 in increased profit. This is about a 1% increase in profits.

The students will be given a set of products (from a real-world dataset) in which they are to apply SAP's TPO package as well as the other advanced analytics and visualization techniques and tools they learned in the class:

- to develop a plan for the next several periods for their products, and
- to understand the past performance of their products.

They will present their results from their analysis to a panel of experts. The setting for the presentation will be that they are the manufacturer and they want to work with and convince the retailer, the panel of experts, to have their products promoted and available on the retailer's shelf with a positive ROI for both.

The two above EL experience examples with analytics are largely from the statistics discipline. Similar EL experiences could/should be developed for the IT and quantitative methods disciplines to enhance the students' learning experience in those BA areas.

## CONCLUSIONS

New undergraduate BI/BA majors and minors, Master of Science degrees, certificates, and concentrations within MBA programs are appearing in business schools. The landscape of courses and topics covered vary as much as the variation of definitions of BI and BA. Regardless of your definition of BI and BA, both include three disciplines: information technology, statistics, and quantitative methods. Most programs emphasize only one or two of these disciplines. We believe all BI or BA program should include a foundational understanding of all three disciplines. These programs should further provide a fundamental competency in understanding business and communication skills. Most undergraduate and MBA programs already provide these foundation courses, while most Master's programs do not.

We suggest the following additionally recommendations:

- For all business students enrolled in undergraduate or MBA programs: Analytics should be integrated across the entire undergraduate and MBA curriculum. If this integration is to occur, a significant paradigm shift must occur in which analytics is promoted and utilized in all business discipline areas. This paradigm shift is similar to what is happening in industry—analytical skills are becoming an essential set of fundamental capabilities of all employees/business students.
- For business school BI/BA majors or minors and for Master's programs that in all likelihood include a course on data mining: Besides providing the students with an introduction to the tools, techniques, and technology, we must emphasize the critical skill of problem solving/modeling. We need to teach our students the craft of problem solving/modeling. The flip classrooms and dataset projects are significant steps to improving their problem-solving/modeling skills. Even so, a gap still remains between the techniques, problem-solving/modeling skills, and real-world decision-making situations. This critical skill can be accomplished by using a problem-

based learning approach or by experiential learning experiences, through, for example, live cases.

- For data mining courses: The number of techniques covered should be fewer (or another class should be added). With this additional time, more time and emphasis should be placed on developing students' problem-solving/models skills by using flip teaching or dataset projects and, to greater extent, specifically adding experiential learning experiences.

"The time has come today."[5]

## NOTES

1. Actually, it was Hans Pete Luhn who first coined the term business intelligence in a 1958 paper.
2. The term business analytics dates back to Fredrick Taylor's time management exercises in the late 19th century.)
3. Herschel use this phrase to define BI. We extended it to also BA.
4. Especially for logistic regression and perhaps some other techniques, they could be considered a data mining technique.
5. Title of the 1968 song by the Chambers Brothers.

## REFERENCES

AACSB International. (2005). *Accreditation standards.* Retrieved from http://www.aacsb.edu/en/accreditation/standards/2003-business/aol/curriculum-management/

Collins, A., Brown, J. S., & Newman, S. E. (1989). Cognitive apprenticeship: Teaching the crafts of reading, writing, and mathematics. In L. B. Resnick (Ed.), *Knowing, learning, and instruction: Essays in honor of Robert Glaser* (pp. 453–494). Hillsdale, NJ: Lawrence Erlbaum Associates.

Committee on the Next Decade in Operations Research (CONDOR). (1988). Operations research: The next decade. *Operations Research, 36*(4), 619–637.

Davenport, T. H., & Harris, J. G. (2007). *Competing on analytics: The new science of winning.* Cambridge, MA: Harvard.

Evans, P. (2010, April). Business intelligence is a growing field. *DataBase Journal.* Retreieved from http://www.databasejournal.com/features/article.php/3878566/Business-Intelligence-is-a-Growing-Field.htm

Gentry, J. W. (1990). *What is experiential learning?, Guide to business gaming and experiential learning.* Retrieved from http://www.wmich.edu/casp/servicelearning/files/What%20is%20Experiential%20Learning.pdf

Gorman, M. F., & Klimberg, R. K. (2014). Benchmarking academic programs in business analytics. *Interfaces, 44*(3), 329–341.

Grossman, T. A. (2002). The Keys to the Vault. *OR/MS Today*, December. Retreived from http://www.lionhrtpub.com/orms/orms-12-02/freducation.html

Herschel, R. (2010, June 1). What is business intelligence? *BeyeNetwork*. Retrieved from http://www.b-eye-network.com/view/13768

INFORMS. (2013a). What is analytics? Retrieved from https://www.informs.org/About-INFORMS/What-is-Analytics

INFORMS. (2013b). Analytics section: Overview. Retrieved from https://www.informs.org/Community/Analytics

Klimberg, R. K., & Cohen, R. M. (1999). Experimental evaluation of a graphical display system to visualizing multiple criteria solutions. *European Journal of Operational Research, 119*, 191–208.

Klimberg, R., & McCullough, B. D. (2013a). Business analytics—Today's green?, *Contemporary Perspectives in Data Mining, 1*, 47–60.

Klimberg, R. K., & McCullough, B. D. (2013b). *Fundamentals of predictive analytics with JMP.* Cary, NC: SAS Press.

Klimberg, R. K., & Miori, V. (2010). Back in business. *ORMS Today, 37*(5), 22–27.

Knowles, M. S. (1990). *The adult learner: A neglected species* (4th ed.). Houston, TX: Gulf Publishing.

Larkin, J., McDermott, J., Simon, D. P., & Simon, H. A. (1980). Expert and novice performance in solving physics problems. *Science, 208*, 1335–1342.

Lomas, N. (2008, November 19). CNET interview. Retrieved from http://news.cnet.com/8301-11386_3-10102273-76.html

Manyika, J., Chui, M., Brown, B., Bughin, J., Dobbs, R., Roxburgh, C., & Hung Byers, A. (2011). *Big data: The next frontier for innovation competition and productivity.* London, UK: McKinsey Global Institute.

Morris, W. (1967). On the art of modeling. *Management Science, 13*, B707–B717.

Powell, S. G. (1995a). Teaching the art of modeling to MBA students. *Interfaces, 25*(3), 88–94.

Powell, S. G. (1995b). Six key modeling heuristics. *Interfaces, 25*(4), 114–125.

Powell, S. G. (2001). Teaching modeling in management science. *INFORMS Transactions on Education, 1*(2), 62–67.

Powell, S. G., & Willemain, T. R. (2003). Model formulation by novices. NSF Proposal, August 15, Decision, Risk and Management.

Schön, D. R. (1983). *The reflective practitioner: How professionals think in action.* New York, NY: Basic Books.

Sillup, G., Klimberg, R., & McSweeney, D. P. (2010, March). Data-driven decision making for new drugs—A collaborative learning experience. *International Journal of Business Intelligence Research,* 1(2), 42–59.

University of California Davis. (2011). 5-step experiential learning cycle definitions. Retrieved from http://www.experientiallearning.ucdavis.edu/module1/el1_40-5step-definitions.pdf

Vessey, I. (1991). Cognitive fit: A theory-based analysis of the graphs versus tables literature. *Decision Sciences, 22*(2), 219–240.

Willemain, T. R. (1994). Insights on modeling from a dozen experts. *Operations Research, 42*(2), 213–222.

Willemain, T. R. (1995). Model formulation: What experts think about and when. *Operations Research, 43*(6), 916–933.

Wixom, B., Ariyachandra, T., Goul, M., Gray, P., Kulkarni, U., & Phillips-Wren, G. (2011). The current state of business intelligence in academia. *Communications of the Association for Information Systems, 29*, Article 16. Available at: http://aisel.aisnet.org/cais/vol29/iss1/16

CHAPTER 6

# MEASURING THE MARKET EFFICIENCY OF CHINESE AUTOMOBILE INDUSTRY BY USING A MAX–MIN DEA MODEL

**Feng Yang, Hangting Hu, Chenchen Yang**
***University of Science and Technology of China***

**Zhimin Huang**
***Adelphi University and***
***Beijing Institute of Technology***

## ABSTRACT

The automobile industry plays a fundamental role in the economic growth of China, and enhancing the sales of the automobile market is important to the development of the Chinese automobile industry. Aiming to better cognize the automobile market, the current chapter tries to discover what factors influence the sales of automobiles in China, and whether the allocation efficiency of the Chinese automobile market is injured by the existing industry policy. We collect 100 representative Chinese vehicle types as

*Contemporary Perspectives in Data Mining, Volume 2*, pages 97–121

samples and gather data about the operational management of those automobile manufacturers. With the help of the efficiency market hypothesis, two different functions (linear and exponential) are proposed to determine the functional relationship between sales and the influencing factors such as price, quality, sale service satisfaction index, and after-sale service satisfaction index. We use a max–min programming format of DEA technique to maximize the efficiency scores of the sales allocation and determine which function form is suitable to characterize Chinese auto sales. The data analysis helps to get the major influencing factor and unearth why the efficiency market hypothesis does not stand in the Chinese automobile industry. We also analyze the efficiency of Chinese automobile market from the perspective of the various brands, and discuss why the varying scale or shareholder structure of automobile manufacturers leads to the sales allocation efficiency to distribute variously.

## INTRODUCTION

The automobile industry plays an extremely vital role to the current economic growth of China. In addition, the Chinese government considers the automotive industry as a future growth engine of the Chinese national economy (State Council of China, 1986). In recent years, with the rapid development of the Chinese economy and the rising of people's wealth, auto sales in China have increased quickly. According to the Annual Report of China Association of Automobile Manufacturers, about 19 million automobiles were sold in China in 2012, which ranks the first among all countries. Notwithstanding, there is a big gap between China and other developed countries with regard to manufacturing competition, technology, business management, and market efficiency in the automobile industry. Additionally, the car consumption environment (the pressure of petroleum supply, city transportation, environmental protection, etc.) is becoming complicated in China. The situations limit the long-term growth of auto sales and hamper the development of the Chinese automobile industry. Considering that the situation cannot be improved in the short term, automobile manufacturers and the government look to improve auto market efficiency and enhance auto sales. As a result, the continuous pursuit of higher automobile sales and a more efficient market has drawn considerable attention. The current chapter aims to study these questions from the consumer's perspective and analyze what factors specifically affect the sales and to what extent they influence the sales.

Tian (2007) considered three factors—real labor, material, and capital inputs—to estimate the production function and illustrated that these intermediate inputs pushed productivity in the Chinese auto industry. Both Oh, Lee, Hwang, and Heshmati (2010) and Choi and Oh (2010) studied product efficiency from the viewpoint of consumers in the Korean auto market. Oh et al. (2010) first estimated the consumers' utility function considering the auto's price and a matrix of quality attributes (maintenance cost, gas station, fuel type, horsepower, fuel cost, etc.). Most of the above quality attributes are quantitative indexes. Choi and Oh (2010) evaluated the product efficiency of Korean hybrid vehicles on the basis of the model developed in Oh et al. (2010). However, Saranga (2009) analyzed the operational efficiency of India's auto component industry in the view of a component supplier and an original equipment manufacturer. Wang, Liu, Wang, and Xie (2008) evaluated customer satisfaction in Chinese automotive after-sales service and considered it as a dependent variable. However, in the current chapter, we regard it as an independent variable and evaluate auto market efficiency based on obtaining the fitting function of auto sales and efficiency market hypothesis.

According to efficiency market hypothesis (EMH) theory, proposed by Fama (1970), if the price fully reflects all available information in a securities market, the market is called an efficient market. Such a definition requires three strict conditions. First, every person is a rational economic man in the securities market. Second, the prices of traded assets (stocks, bonds, or property) reflect the balance of supply and demand. Third, the share price fully reflects all public available information on the traded assets in a strong efficient market, which is called "effective information." On the basis of previous theoretical research, Jensen (1978) stated his belief that the proposition of the efficient market hypothesis had numerous pieces of solid empirical evidence supporting it compared to other propositions in economics. In addition, Palan (2004) declared that the significance of EMH to modern finance could never be ignored.

The concept of an efficient market was initially proposed to explain the securities market and later introduced to discuss other industrial markets. For example, in an efficient commodities exchange market, some actual information about the products such as price, quality, and service are better reflected by its sales volume. Besides, there exists a one-to-one correspondence between the price, quality, service of the product, and sales. In other words, excellent products should have better sales than poor products. EMH has been prevalently discussed and verified in many studies. For example, Case and Shiller (1989) and Guntermann and Norrbin (1991) studied the efficiency of the housing market using EMH. Yang, Yang, Xia, and Ang (2012) applied EMH in the Chinese shampoo industry. In this study, we also apply the concept of EMH to measure the degree of market efficiency of the Chinese automobile industry and analyze whether the efficiency market hypothesis stands there.

We use mass statistical data to measure the market efficiency of the Chinese automobile industry by employing data envelopment analysis (DEA) as a tool. DEA , first developed by Charnes, Cooper, and Rhodes (1978), is a nonparametric method for measuring the efficiency of decision making units (DMUs) such as public sectors, hospitals, firms, universities, banks, and the like (Banker, Das, & Datar, 1989; Charnes, Clarke, Cooper, & Golany, 1985; Cook & Green, 2004; Emrouznejad, Parker, & Tavares, 2008; Ertay, Ruan, & Tuzkaya, 2006; Seiford & Zhu, 1999). DEA has been regarded as a valid analytical method and a practical decision support (Ray, 2004) for, without requiring a complete specific form of the production function, it just requires a general form of the production function. In this chapter, we utilize the max–min DEA model, which is proposed by maximizing the minimum efficiency score of all units, to estimate the auto market efficiency and specify the auto sales function in China.

The rest of the chapter unfolds as follows. We report the background of the Chinese automobile industry and display the related data in Section 2. Subsequently, we make two different assumptions about the form of the automobile sales function, and introduce the max–min DEA model in Section 3. Section 4 presents the results and discussions. Finally, we illustrate the managerial implications of our study in the last section.

## PROBLEM DESCRIPTIONS

The automobile industry is regarded as a leading pillar of industry in the Chinese national economy. Compared with other industries in China, the automotive market has some special characteristics. First, the auto industry has high capital intensity, depends on foreign technology excessively, and enjoys high profitability. Tian (2007) found that automobile companies were smaller and less competitive compared with computer firms. Second, considering that automobile is an expensive, functional, and long-lasting purchase, consumers usually spend a good deal of time and budget on their automobile purchasing decision. Before buying a car, a consumer might make a car budget and search the relevant information on brands, types, sale price, and service experience from newspapers, internet, auto magazines, televisions, and so on. Third, the Chinese government regards the automobile industry as a leading pillar industry of the Chinese national economy and has made a lot of intervening policies to inspire this industry.

Manufacturers care a lot about what factors influence consumers' decisions to buy a car. Figure 6.1 reports the major influencing factors on auto sales obtained from a questionnaire by cheshi.com, the top online car market in China. Obviously, price and quality are more important than others. We gather that the sale service and after-sale service have also aroused great concerns. Though there are many other factors influencing automotive sales, such as convenient location, large-scale operation, environmental

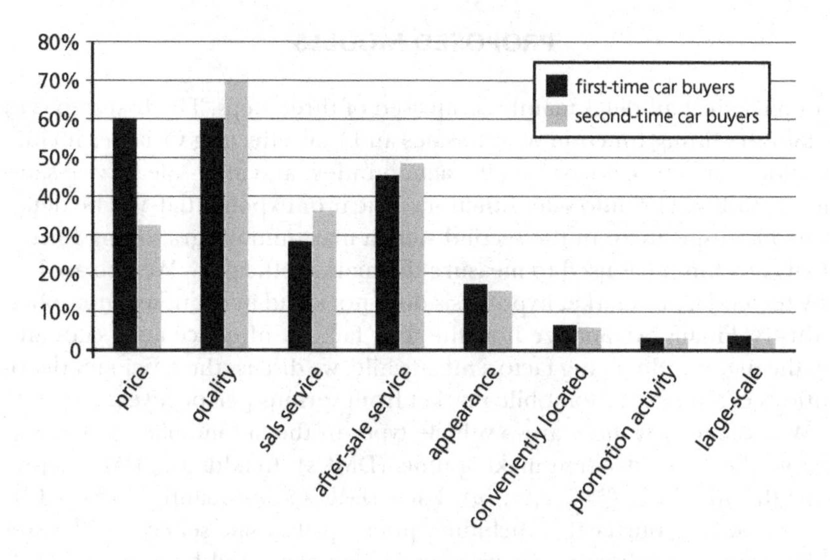


**Figure 6.1** The influencing factors on auto sales (*Source:* www.cheshi.com).

friendliness, promotion activity, appearance, quick extraction of car, and so on, we can see that these additional factors are not so important from the histogram below; thus, we do not consider these factors in this study. In conclusion, we choose price, car quality, sale service satisfaction index, and after-sale service satisfaction indices as the primary influencing factors on auto sales.

We gather a set of 100 representative vehicle types, which belong to different firms or different brands in the Chinese market for the year 2012, and the detailed data are given in Table A1 in the Appendix. For each vehicle type, the manufacturer, brand, price, quality, sale service satisfaction index (SSI), and after-sale service satisfaction index (ASI) are reported in columns 3–8 of Table A1. The sales data are also given in the rightmost column. The data of price and quality in columns 5 and 6 come from www.autohome.com.cn, which is the most frequently visited car website all over the world. The data of sale service satisfaction index and after-sale service satisfaction index are obtained from Lansion, which is the first independent consulting organization that focuses on investigating satisfaction in China. Additionally, the data of auto sales are from China Association of Automobile Manufacturers (2012).

In order to specify auto sales function in the efficiency calculations, price, quality, sale service satisfaction index, and after-sale service satisfaction index are taken as input variables and sales is considered as output variable. Next we show how to determine the auto sales function in a max–min DEA model.

## PROPOSED MODELS

Our analytical model is mainly comprised of three steps. The first step is to obtain the fitting function of auto sales and four effecting variables including price, quality, sale service satisfaction index, and after-sale service satisfaction index. The auto sales functions, linear or exponential, will be determined and specified. In the second step, a max–min programming format of DEA technique is used to measure the market efficiency. We also analyze why the efficiency market hypothesis does not stand in Chinese automobile industry. Finally, we analyze how the four factors influence auto sales and get the major influencing factor; meanwhile, we discuss the efficiency distributions of Chinese automobile market from various perspectives.

We assume that there are $n$ vehicle types in the automobile market and denote them as $n$ decision making units (DMUs). In addition, $DMU_j$ represents the $j$th DMU $(j = 1, 2, \ldots, n)$. Each DMU's sales quantity, denoted by $y_j$, is swayed by four factors including price, quality, sale service satisfaction index, and after-sale service satisfaction index expressed by $x_{ij}$ $(i = 1, 2, 3, 4;$ $j = 1, 2, \ldots, n)$. We define sales amount as a function on four factors as follows:

$$S = f(P, Q, SSI, ASI) \quad (6.1)$$

In the Equation 6.1, $S$ represents car sales and $P$, $Q$, $SSI$, and $ASI$ are short for price, quality, sale service satisfaction index, and after-sale service satisfaction index, respectively. Additionally, we define the relationship between sales and these four variables as a function $f()$. As we do not know the concrete form of the sales function, thereafter, we will make two assumptions of the form of automobile sales function: linear and exponential forms.

### Linear Sales Function

**Hypothesis 1:** *The automotive market conforms to the efficient market hypothesis.*

**Hypothesis 2:** *The form of automotive sales function is linear.*

According to the two hypotheses, we obtain the form of sales function as follows:

$$S = f(P, Q, SSI, ASI) = v_1 \times P + v_2 \times Q + v_3 \times SSI + v_4 \times ASI \quad (6.2)$$

where the coefficients $v_i (i = 1,2,3,4)$ denote the influencing elasticity of each factor to the auto sales.

For better presentation, using $x_j$ to replace the influencing factors, we rewrite Equation 6.2 as follows:

$$S_j = f(x_1, x_2, x_3, x_4) = \sum_{i=1}^{4} v_i x_{ij} \tag{6.3}$$

$S_j$ denotes the maximum feasible quantity of the $j$th type automobile in the market under the attribution of four variables. It is a theoretical value that builds on the basis of the efficient market hypothesis.

To represent the sales allocation efficiency, we use the ratio of actual auto sales, denoted by $y$, to the theoretical maximum auto sales, denoted by $S$, with the expression as follows:

$$\hat{\theta} = \frac{y}{S} \tag{6.4}$$

$\hat{\theta}$ denotes the sales allocation efficiency.

After substituting (6.3) in (6.4), we obtain the sales allocation efficiency of DMU$j$:

$$\hat{\theta}_j = \frac{y_j}{\sum_{i=1}^{4} v_i x_{ij}} \tag{6.5}$$

Actual auto sales of each DMU$j$ should equal to the theoretical maximum auto sales of each DMU$j$ for any $j = 1, 2, \ldots, n$ in the market if the market is efficient. If it happens, every DMU$j$'s efficiency score should equal to one theoretically, as shown in (6.6):

$$\hat{\theta}_j = \frac{y_j}{\sum_{i=1}^{4} v_i x_{ij}} = 1 \quad j = 1, 2, \cdots, n \tag{6.6}$$

From above analysis, $n$ linear equations are acquired. Subsequently, we need to solve the equation set grouping of $n$ linear equations for the sake of acquiring the auto sales function and the influence extent of each factor. Nevertheless, there are generally no feasible solutions to the equation group above. Data envelopment analysis has established itself as an extraordinarily useful tool in assessing efficiencies of decision making units in recent years. Here, using publicly available data, we introduce the max–min DEA model to force the sales allocation efficiency values of all DMUs to approach to one as close as possible:

$$\begin{aligned} &\max \ \min \ \theta_j \\ &\text{subject to} \quad \theta_j = \frac{y_j}{\sum_{i=1}^{4} v_i x_{ij}} \le 1, \qquad j = 1, \ldots, n \\ &\qquad\qquad\quad v_i \ge 0, \qquad\qquad\qquad\quad i = 1, \ldots, 4 \end{aligned} \tag{6.7}$$

The first constraint implies that no DMU may have an efficiency value greater than 100% with the same weights $v_i$. The main purpose of the above programming is to find a maximum from the minimal value of a bunch of efficiency scores. It guarantees that all of the DMUs are closest to full efficiency. As there exist measurement errors, it is easy to be accepted that θ approaches closely to 1 if $\theta_j \geq 0.95$. Otherwise, if θ is too small, it is said that the two hypotheses are not proved.

It is noticed that the variables $v_i$ in Model (6.7) are the coefficients to illustrate the relationship function. While in the view of linear programming model, they are the identical weights that are the optimal solutions of the linear programming model. Thus we obtain the common weights $v_i$ by solving Model (6.7).

In order to calculate conveniently, we transform Model (6.7) into the following linear programming:

$$\begin{aligned} &\max\ \theta \\ &s.t.\quad \theta_j = \frac{y_j}{\sum_{i=1}^{4} v_i x_{ij}} \leq 1, \quad j = 1,\ldots,n, \\ &\qquad\ \theta_j \geq \theta, \qquad\qquad\qquad j = 1,\ldots,n, \\ &\qquad\ v_i \geq 0, \qquad\qquad\qquad i = 1,\ldots 4. \end{aligned} \tag{6.8}$$

As one of the constraints of the Model (6.8) is nonlinear, it is not convenient to solve. By appropriate conversion, the nonlinear constraints are transformed into integral forms as follows:

$$\begin{aligned} &\max\ \theta \\ &s.t.\quad y_j - \sum_{i=1}^{4} v_i x_{ij} \leq 0, \qquad j = 1,\ldots,n, \\ &\qquad\ \theta \times \sum_{i=1}^{4} v_i x_{ij} - y_j \leq 0, \quad j = 1,\ldots,n, \\ &\qquad\ v_i \geq 0, \qquad\qquad\qquad i = 1,\ldots 4, \\ &\qquad\ \theta \geq 0. \end{aligned} \tag{6.9}$$

It is reasonable to view Model (6.9) as a parametric linear programming with parameter θ. On the premise of satisfying the constraints in model (6.9), the feasible solutions are attained by presetting a small initial value $\theta_0$ for θ and increasing θ in the light of $\theta_t = \theta_0 + \varepsilon \times t$ for every step $t$, where θ is small and $\theta \geq 0$. We substitute $\theta_0$ into the above linear programming and find the feasible solutions, then repeat the procedure according to $\theta_t = \theta_0 + \varepsilon \times t$ until feasible solutions do not exist. Finally, we acquire the maximum $\theta_t$ and the optimal solutions of Model (6.9).

## Exponential Sales Function

**Hypothesis 3:** *The automotive market follows the efficient market hypothesis.*

**Hypothesis 4:** *The sales function in the automotive market is an exponential function.*

Similar in form to the distinguished Cobb-Douglas production function, the form of auto sales function in Hypothesis 4 is given as below:

$$S = f(P, Q, SSI, ASI) = P^{v_1} \times Q^{v_2} \times SSI^{v_3} \times ASI^{v_4} \tag{6.10}$$

By analogy to Section 3.1, we obtain the theoretical sales and sales allocation efficiency of DMU$j$ as the following expression:

$$S_j = f(x_1, x_2, x_3, x_4) = \prod_{i=1}^{4} x_{ij}^{v_i} \tag{6.11}$$

$$\hat{\theta}_j = \frac{y_j}{\prod_{i=1}^{4} x_{ij}^{v_i}} \tag{6.12}$$

If the auto market is efficient, there exist $n$ expressions as follows:

$$\hat{\theta}_j = \frac{y_j}{\prod_{i=1}^{4} x_{ij}^{v_i}} = 1 \quad j = 1, 2, \cdots, n \tag{6.13}$$

The sales function is acquired by solving the mathematical nonprogramming problem as below:

$$\begin{aligned} & \max \ \min \ \theta_j \\ & s.t. \ \theta_j = \frac{y_j}{\prod_{i=1}^{4} x_{ij}^{v_i}} \le 1, \ j = 1, \ldots, n, \\ & v_i \ge 0, \qquad\qquad i = 1, \ldots, 4. \end{aligned} \tag{6.14}$$

For the sake of computation simplicity, we let $\tilde{y}_j = \ln y_j$, $\tilde{x}_{ij} = \ln x_{ij}$, $\tilde{\theta}_j = \ln \theta_j$, then

$$\theta_j = \frac{y_j}{\prod_{i=1}^{4} x_{ij}^{v_i}} \le 1$$

is transformed to

$$\ln \theta_j = \ln y_j - \ln \prod_{i=1}^{4} x_{ij}^{v_i} \le 0 \text{ or } \tilde{\theta}_j = \tilde{y}_j - \sum_{i=1}^{4} v_i \tilde{x}_{ij} \le 0.$$

Thus, based on the above transformation, we turn the first constraint of Model (6.14) into two constraints—that is,

$$\tilde{y}_j - \sum_{i=1}^{4} v_i \tilde{x}_{ij} \le 0 \text{ and } \tilde{\theta} + \sum_{i=1}^{4} v_i \tilde{x}_{ij} - \tilde{y}_j \le 0.$$

Similar to the above section of linear sales function assumption, $\min \tilde{\theta}_j$ is redesigned to $\tilde{\theta}_j \ge \tilde{\theta}$. Hence, the feasible solutions are acquired by using the parametric linear programming below instead of Model (6.14):

$$\begin{aligned} & \max\ \tilde{\theta} \\ & s.t.\ \ \tilde{y}_j - \sum_{i=1}^{4} v_i \tilde{x}_{ij} \le 0, \qquad j = 1,\ldots,n, \\ & \tilde{\theta} + \sum_{i=1}^{4} v_i \tilde{x}_{ij} - \tilde{y}_j \le 0, \quad j = 1,\ldots,n, \\ & v_i \ge 0, \qquad i = 1,\ldots,4. \end{aligned} \tag{6.15}$$

If Model (6.15) has feasible solutions satisfying $\tilde{\theta} = 0$, it is equivalent to Expressions (6.13). The optimal weights $v_i (i = 1,2,\ldots,4)$ and efficiency value $\tilde{\theta}$ are determined by solving Model (6.15). After that we clarify the specific auto sales function expression.

## RESULTS AND DISCUSSIONS

The market efficiency analysis of China's automobile industry, solving by max–min DEA-based models, will be given in this section. As we know that the relationship of price and sales is negative generally, we get the positive relationship of transformed price and the desired values by converting $p$ to $p'$ through the formula $p' = M - p$. Hence, according to general rules, these variables including price, quality, sale service satisfaction index, and after-sale service satisfaction index are rising with the amount of sales. As price and sales are quantitative data while the others are qualitative data, the coefficients of them cannot be compared due to disparate orders of magnitude. Here we utilize $x'_{ij} = x_{ij} / \max_i \{x_{ij}\}$ to normalize the inputs and outputs within 0–1 range. Then we show the modified data of all DMUs used in model (6.9) and (6.15) in Table 6.1 and efficiency values in Tables 6.2 and 6.3.

It is easy for us to see that the efficiency values in Table 6.2 are higher than the efficiency scores in Table 6.1, and the latter are close to 1. Thus we conclude that it is better to use the exponential auto sales function to describe the relationship of the mentioned variables to the automotive sales.

The coefficients of the price, quality, SSI, and ASI are not the same due to the assumption of the form of the sales function are different from

**TABLE 6.1 The Modified Data and Results by Model (6.9)**

| No. | Price | Quality | SSI | ASI | Sales | Efficiency values |
|---|---|---|---|---|---|---|
| 1 | 0.8532 | 0.9002 | 0.9988 | 0.9954 | 0.0114 | 0.0114 |
| 2 | 0.9085 | 0.8764 | 0.9988 | 0.9954 | 0.2876 | 0.2798 |
| 3 | 0.7103 | 0.9588 | 0.9988 | 0.9954 | 0.0604 | 0.0651 |
| 4 | 0.8456 | 0.9566 | 0.9988 | 0.9954 | 0.7212 | 0.7161 |
| 5 | 0.9019 | 0.8850 | 0.9988 | 0.9954 | 0.3591 | 0.3501 |
| 6 | 0.8465 | 0.8720 | 0.9988 | 0.9954 | 0.2301 | 0.2324 |
| 7 | 0.8574 | 0.8894 | 0.9988 | 0.9954 | 0.7649 | 0.7647 |
| 8 | 0.8716 | 0.8850 | 0.9107 | 0.9166 | 0.1679 | 0.1709 |
| 9 | 0.9404 | 0.8937 | 0.9107 | 0.9166 | 0.4412 | 0.4304 |
| 10 | 0.9412 | 0.9826 | 0.9107 | 0.9166 | 0.1143 | 0.1095 |
| 11 | 0.9211 | 0.9436 | 0.9107 | 0.9166 | 0.0115 | 0.0112 |
| 12 | 0.9479 | 0.9262 | 0.9107 | 0.9166 | 0.1468 | 0.1416 |
| 13 | 0.9563 | 0.7852 | 0.9107 | 0.9166 | 0.2784 | 0.2750 |
| 14 | 0.8927 | 0.9067 | 0.9107 | 0.9166 | 0.1382 | 0.1383 |
| 15 | 0.8385 | 0.9458 | 0.9952 | 0.9930 | 0.1724 | 0.1724 |
| 16 | 0.8960 | 0.9349 | 0.9952 | 0.9930 | 0.0240 | 0.0233 |
| 17 | 0.8346 | 0.8829 | 0.9952 | 0.9930 | 0.5155 | 0.5236 |
| 18 | 0.5987 | 0.9566 | 0.9952 | 0.9930 | 0.3220 | 0.3746 |
| 19 | 0.8898 | 0.9696 | 0.9952 | 0.9930 | 0.4784 | 0.4621 |
| 20 | 0.9236 | 0.9024 | 0.9952 | 0.9930 | 0.4927 | 0.4731 |
| 21 | 0.9068 | 0.9393 | 0.8952 | 0.9027 | 0.0960 | 0.0950 |
| 22 | 0.8742 | 0.9284 | 0.8952 | 0.9027 | 0.1886 | 0.1908 |
| 23 | 0.7163 | 0.9154 | 0.8952 | 0.9027 | 0.0328 | 0.0368 |
| 24 | 0.8734 | 0.9262 | 0.8952 | 0.9027 | 0.2895 | 0.2931 |
| 25 | 0.9068 | 0.9284 | 0.8952 | 0.9027 | 0.4466 | 0.4431 |
| 26 | 0.8658 | 0.8720 | 0.8952 | 0.9027 | 0.1268 | 0.1305 |
| 27 | 0.9137 | 0.8200 | 0.8952 | 0.9027 | 0.0387 | 0.0391 |
| 28 | 0.6500 | 0.8764 | 0.9369 | 0.9096 | 0.1284 | 0.1510 |
| 29 | 0.8927 | 0.9262 | 0.9369 | 0.9096 | 0.1213 | 0.1207 |
| 30 | 0.4820 | 0.9870 | 0.9369 | 0.9096 | 0.0879 | 0.1136 |
| 31 | 0.8448 | 0.9718 | 0.9369 | 0.9096 | 0.3464 | 0.3513 |
| 32 | 0.8381 | 0.9783 | 0.9369 | 0.9096 | 0.3722 | 0.3785 |
| 33 | 0.9716 | 0.7939 | 0.8679 | 0.9340 | 0.1791 | 0.1753 |
| 34 | 0.9867 | 0.9566 | 0.8679 | 0.9340 | 0.1514 | 0.1422 |
| 35 | 0.8574 | 0.9176 | 0.8679 | 0.9340 | 0.1288 | 0.1315 |
| 36 | 0.9498 | 0.7874 | 0.8679 | 0.9340 | 0.0493 | 0.0489 |
| 37 | 0.9857 | 0.9241 | 0.9738 | 0.9258 | 0.0654 | 0.0613 |
| 38 | 0.9389 | 0.8915 | 0.9738 | 0.9258 | 0.0640 | 0.0620 |
| 39 | 0.9699 | 0.9024 | 0.9738 | 0.9258 | 0.2296 | 0.2180 |
| 40 | 0.9565 | 0.9544 | 0.9738 | 0.9258 | 0.5147 | 0.4874 |
| 41 | 0.9807 | 0.8547 | 0.9738 | 0.9258 | 0.0852 | 0.0812 |

*(continued)*

**TABLE 6.1 The Modified Data and Results by Model (6.9) (continued)**

| No. | Price | Quality | SSI | ASI | Sales | Efficiency values |
|---|---|---|---|---|---|---|
| 42 | 0.9120 | 0.8980 | 0.9738 | 0.9258 | 0.0555 | 0.0545 |
| 43 | 0.9111 | 0.8698 | 0.9738 | 0.9258 | 0.1185 | 0.1171 |
| 44 | 0.8658 | 0.8807 | 0.9738 | 0.9258 | 0.0682 | 0.0691 |
| 45 | 0.8296 | 0.9479 | 0.9893 | 1.0000 | 0.8818 | 0.8856 |
| 46 | 0.6340 | 0.9675 | 0.9893 | 1.0000 | 0.8341 | 0.9441 |
| 47 | 0.7807 | 0.9132 | 0.9893 | 1.0000 | 0.3729 | 0.3886 |
| 48 | 0.8547 | 0.8915 | 0.9893 | 1.0000 | 0.3521 | 0.3525 |
| 49 | 0.8850 | 0.9262 | 0.9893 | 1.0000 | 0.4260 | 0.4160 |
| 50 | 0.8969 | 0.8915 | 0.9512 | 0.9409 | 0.1573 | 0.1560 |
| 51 | 0.8968 | 0.8829 | 0.9512 | 0.9409 | 0.1729 | 0.1718 |
| 52 | 0.7111 | 0.9653 | 0.9512 | 0.9409 | 0.4919 | 0.5388 |
| 53 | 0.8786 | 0.9219 | 0.9821 | 0.9722 | 0.2659 | 0.2625 |
| 54 | 0.6330 | 1.0000 | 0.9821 | 0.9722 | 0.1379 | 0.1561 |
| 55 | 0.6861 | 0.9566 | 0.9821 | 0.9722 | 0.0175 | 0.0193 |
| 56 | 0.8816 | 0.9371 | 0.9702 | 0.9873 | 0.3054 | 0.2995 |
| 57 | 0.9460 | 0.8633 | 0.9702 | 0.9873 | 0.3078 | 0.2952 |
| 58 | 0.8346 | 0.9306 | 0.9702 | 0.9873 | 0.9905 | 1.0000 |
| 59 | 0.7939 | 0.9197 | 0.9702 | 0.9873 | 0.4560 | 0.4730 |
| 60 | 0.5591 | 0.9610 | 0.9810 | 0.9618 | 0.0664 | 0.0801 |
| 61 | 0.6582 | 0.9197 | 0.9810 | 0.9618 | 0.8315 | 0.9454 |
| 62 | 0.8952 | 0.9176 | 0.9810 | 0.9618 | 1.0000 | 0.9805 |
| 63 | 0.7279 | 0.9653 | 0.9810 | 0.9618 | 0.2024 | 0.2175 |
| 64 | 0.8035 | 0.8742 | 0.9810 | 0.9618 | 0.1856 | 0.1940 |
| 65 | 0.8885 | 0.9197 | 1.0000 | 0.9687 | 0.4452 | 0.4365 |
| 66 | 0.3552 | 0.9805 | 1.0000 | 0.9687 | 0.0910 | 0.1272 |
| 67 | 0.8213 | 0.9284 | 1.0000 | 0.9687 | 0.5430 | 0.5531 |
| 68 | 0.0275 | 0.9761 | 1.0000 | 0.9687 | 0.0669 | 0.1287 |
| 69 | 0.6197 | 0.8742 | 1.0000 | 0.9687 | 0.2139 | 0.2514 |
| 70 | 0.6768 | 0.9328 | 1.0000 | 0.9687 | 0.0326 | 0.0364 |
| 71 | 0.9882 | 0.8698 | 0.9690 | 0.9583 | 0.2383 | 0.2241 |
| 72 | 0.9546 | 0.8677 | 0.9690 | 0.9583 | 0.1747 | 0.1675 |
| 73 | 0.9588 | 0.8482 | 0.9690 | 0.9583 | 0.1632 | 0.1567 |
| 74 | 0.9177 | 0.9610 | 0.9690 | 0.9583 | 0.1484 | 0.1427 |
| 75 | 0.9456 | 0.8807 | 0.9690 | 0.9583 | 0.0302 | 0.0290 |
| 76 | 0.9177 | 0.9610 | 0.9690 | 0.9583 | 0.1438 | 0.1382 |
| 77 | 0.9152 | 0.9523 | 0.9512 | 0.9282 | 0.0436 | 0.0424 |
| 78 | 0.9530 | 0.8178 | 0.9512 | 0.9282 | 0.4880 | 0.4767 |
| 79 | 0.8994 | 0.9154 | 0.9512 | 0.9282 | 0.1494 | 0.1476 |
| 80 | 0.9456 | 0.8959 | 0.9512 | 0.9282 | 0.0392 | 0.0379 |
| 81 | 0.9631 | 0.8829 | 0.9321 | 0.9386 | 0.1792 | 0.1718 |
| 82 | 0.9852 | 0.8503 | 0.9321 | 0.9386 | 0.1014 | 0.0966 |

*(continued)*

**TABLE 6.1 The Modified Data and Results by Model (6.9) (continued)**

| No. | Price | Quality | SSI | ASI | Sales | Efficiency values |
|---|---|---|---|---|---|---|
| 83 | 0.9840 | 0.9349 | 0.9321 | 0.9386 | 0.0916 | 0.0859 |
| 84 | 0.9372 | 0.9631 | 0.9321 | 0.9386 | 0.0665 | 0.0637 |
| 85 | 0.9719 | 0.8829 | 0.8369 | 0.8030 | 0.1056 | 0.1044 |
| 86 | 0.9377 | 0.9696 | 0.8369 | 0.8030 | 0.0898 | 0.0890 |
| 87 | 0.9800 | 0.8633 | 0.8369 | 0.8030 | 0.0164 | 0.0162 |
| 88 | 0.9543 | 0.8547 | 0.8369 | 0.8030 | 0.0160 | 0.0161 |
| 89 | 0.9746 | 0.8612 | 0.8500 | 0.8250 | 0.2534 | 0.2499 |
| 90 | 0.9582 | 0.8980 | 0.8500 | 0.8250 | 0.0266 | 0.0263 |
| 91 | 1.0000 | 0.8742 | 0.8500 | 0.8250 | 0.0124 | 0.0120 |
| 92 | 0.9622 | 0.8655 | 0.8500 | 0.8250 | 0.0084 | 0.0083 |
| 93 | 0.7936 | 0.9436 | 0.9655 | 0.9537 | 0.2851 | 0.2965 |
| 94 | 0.8658 | 0.9414 | 0.9655 | 0.9537 | 0.0504 | 0.0502 |
| 95 | 0.6508 | 0.9393 | 0.9655 | 0.9537 | 0.0444 | 0.0507 |
| 96 | 0.6768 | 0.9544 | 0.9655 | 0.9537 | 0.0441 | 0.0493 |
| 97 | 0.6323 | 0.9393 | 0.9750 | 0.9606 | 0.6043 | 0.6968 |
| 98 | 0.8975 | 0.8915 | 0.9750 | 0.9606 | 0.5198 | 0.5121 |
| 99 | 0.7331 | 0.9349 | 0.9750 | 0.9606 | 0.0618 | 0.0667 |
| 100 | 0.4904 | 0.9479 | 0.9750 | 0.9606 | 0.2683 | 0.3419 |

**TABLE 6.2 The Modified Data and Results by Model (6.15)**

| No. | Price | Quality | SSI | ASI | Sales | Efficiency values |
|---|---|---|---|---|---|---|
| 1 | 0.9651 | 0.9312 | 0.9998 | 0.9993 | 0.6435 | 0.6480 |
| 2 | 0.9789 | 0.9137 | 0.9998 | 0.9993 | 0.9006 | 0.9164 |
| 3 | 0.9248 | 0.9725 | 0.9998 | 0.9993 | 0.7762 | 0.7632 |
| 4 | 0.9631 | 0.9710 | 0.9998 | 0.9993 | 0.9739 | 0.9567 |
| 5 | 0.9773 | 0.9201 | 0.9998 | 0.9993 | 0.9183 | 0.9307 |
| 6 | 0.9633 | 0.9104 | 0.9998 | 0.9993 | 0.8828 | 0.9009 |
| 7 | 0.9662 | 0.9233 | 0.9998 | 0.9993 | 0.9786 | 0.9903 |
| 8 | 0.9698 | 0.9201 | 0.9861 | 0.9871 | 0.8577 | 0.8737 |
| 9 | 0.9865 | 0.9265 | 0.9861 | 0.9871 | 0.9348 | 0.9475 |
| 10 | 0.9867 | 0.9886 | 0.9861 | 0.9871 | 0.8270 | 0.8064 |
| 11 | 0.9819 | 0.9620 | 0.9861 | 0.9871 | 0.6441 | 0.6386 |
| 12 | 0.9882 | 0.9499 | 0.9861 | 0.9871 | 0.8470 | 0.8459 |
| 13 | 0.9902 | 0.8418 | 0.9861 | 0.9871 | 0.8980 | 0.9619 |
| 14 | 0.9750 | 0.9359 | 0.9861 | 0.9871 | 0.8422 | 0.8491 |
| 15 | 0.9613 | 0.9635 | 0.9993 | 0.9990 | 0.8598 | 0.8487 |
| 16 | 0.9759 | 0.9560 | 0.9993 | 0.9990 | 0.7027 | 0.6964 |
| 17 | 0.9602 | 0.9185 | 0.9993 | 0.9990 | 0.9472 | 0.9619 |

*(continued)*

**TABLE 6.2 The Modified Data and Results by Model (6.15) (continued)**

| No. | Price | Quality | SSI | ASI | Sales | Efficiency values |
|---|---|---|---|---|---|---|
| 18 | 0.8872 | 0.9710 | 0.9993 | 0.9990 | 0.9096 | 0.8970 |
| 19 | 0.9743 | 0.9798 | 0.9993 | 0.9990 | 0.9412 | 0.9192 |
| 20 | 0.9825 | 0.9328 | 0.9993 | 0.9990 | 0.9436 | 0.9485 |
| 21 | 0.9785 | 0.9590 | 0.9836 | 0.9849 | 0.8131 | 0.8084 |
| 22 | 0.9704 | 0.9514 | 0.9836 | 0.9849 | 0.8670 | 0.8665 |
| 23 | 0.9266 | 0.9422 | 0.9836 | 0.9849 | 0.7276 | 0.7330 |
| 24 | 0.9702 | 0.9499 | 0.9836 | 0.9849 | 0.9012 | 0.9015 |
| 25 | 0.9785 | 0.9514 | 0.9836 | 0.9849 | 0.9357 | 0.9348 |
| 26 | 0.9683 | 0.9104 | 0.9836 | 0.9849 | 0.8353 | 0.8570 |
| 27 | 0.9801 | 0.8701 | 0.9836 | 0.9849 | 0.7407 | 0.7797 |
| 28 | 0.9052 | 0.9137 | 0.9903 | 0.9860 | 0.8363 | 0.8578 |
| 29 | 0.9750 | 0.9499 | 0.9903 | 0.9860 | 0.8318 | 0.8309 |
| 30 | 0.8395 | 0.9914 | 0.9903 | 0.9860 | 0.8061 | 0.7900 |
| 31 | 0.9629 | 0.9813 | 0.9903 | 0.9860 | 0.9155 | 0.8973 |
| 32 | 0.9612 | 0.9857 | 0.9903 | 0.9860 | 0.9212 | 0.9005 |
| 33 | 0.9937 | 0.8490 | 0.9790 | 0.9899 | 0.8629 | 0.9202 |
| 34 | 0.9971 | 0.9710 | 0.9790 | 0.9899 | 0.8495 | 0.8373 |
| 35 | 0.9662 | 0.9437 | 0.9790 | 0.9899 | 0.8366 | 0.8401 |
| 36 | 0.9887 | 0.8436 | 0.9790 | 0.9899 | 0.7600 | 0.8137 |
| 37 | 0.9968 | 0.9484 | 0.9961 | 0.9886 | 0.7825 | 0.7804 |
| 38 | 0.9861 | 0.9249 | 0.9961 | 0.9886 | 0.7808 | 0.7908 |
| 39 | 0.9933 | 0.9328 | 0.9961 | 0.9886 | 0.8827 | 0.8892 |
| 40 | 0.9902 | 0.9695 | 0.9961 | 0.9886 | 0.9470 | 0.9324 |
| 41 | 0.9957 | 0.8973 | 0.9961 | 0.9886 | 0.8036 | 0.8280 |
| 42 | 0.9797 | 0.9297 | 0.9961 | 0.9886 | 0.7695 | 0.7772 |
| 43 | 0.9795 | 0.9088 | 0.9961 | 0.9886 | 0.8300 | 0.8496 |
| 44 | 0.9683 | 0.9169 | 0.9961 | 0.9886 | 0.7859 | 0.8007 |
| 45 | 0.9589 | 0.9650 | 0.9984 | 1.0000 | 0.9900 | 0.9764 |
| 46 | 0.8998 | 0.9784 | 0.9984 | 1.0000 | 0.9855 | 0.9667 |
| 47 | 0.9456 | 0.9406 | 0.9984 | 1.0000 | 0.9213 | 0.9232 |
| 48 | 0.9655 | 0.9249 | 0.9984 | 1.0000 | 0.9168 | 0.9269 |
| 49 | 0.9731 | 0.9499 | 0.9984 | 1.0000 | 0.9320 | 0.9272 |
| 50 | 0.9761 | 0.9249 | 0.9926 | 0.9910 | 0.8525 | 0.8639 |
| 51 | 0.9760 | 0.9185 | 0.9926 | 0.9910 | 0.8601 | 0.8751 |
| 52 | 0.9250 | 0.9769 | 0.9926 | 0.9910 | 0.9434 | 0.9277 |
| 53 | 0.9715 | 0.9468 | 0.9973 | 0.9958 | 0.8944 | 0.8925 |
| 54 | 0.8994 | 1.0000 | 0.9973 | 0.9958 | 0.8420 | 0.8160 |
| 55 | 0.9171 | 0.9710 | 0.9973 | 0.9958 | 0.6776 | 0.6679 |
| 56 | 0.9723 | 0.9575 | 0.9955 | 0.9981 | 0.9054 | 0.8973 |
| 57 | 0.9878 | 0.9039 | 0.9955 | 0.9981 | 0.9060 | 0.9281 |
| 58 | 0.9602 | 0.9529 | 0.9955 | 0.9981 | 0.9992 | 0.9937 |

*(continued)*

**TABLE 6.2 The Modified Data and Results by Model (6.15) (continued)**

| No. | Price | Quality | SSI | ASI | Sales | Efficiency values |
|---|---|---|---|---|---|---|
| 59 | 0.9492 | 0.9453 | 0.9955 | 0.9981 | 0.9374 | 0.9372 |
| 60 | 0.8721 | 0.9740 | 0.9971 | 0.9942 | 0.7838 | 0.7731 |
| 61 | 0.9080 | 0.9453 | 0.9971 | 0.9942 | 0.9853 | 0.9877 |
| 62 | 0.9757 | 0.9437 | 0.9971 | 0.9942 | 1.0000 | 1.0000 |
| 63 | 0.9301 | 0.9769 | 0.9971 | 0.9942 | 0.8726 | 0.8567 |
| 64 | 0.9519 | 0.9120 | 0.9971 | 0.9942 | 0.8657 | 0.8844 |
| 65 | 0.9740 | 0.9453 | 1.0000 | 0.9953 | 0.9355 | 0.9340 |
| 66 | 0.7723 | 0.9871 | 1.0000 | 0.9953 | 0.8089 | 0.7948 |
| 67 | 0.9567 | 0.9514 | 1.0000 | 0.9953 | 0.9513 | 0.9470 |
| 68 | 0.2093 | 0.9842 | 1.0000 | 0.9953 | 0.7843 | 0.7940 |
| 69 | 0.8948 | 0.9120 | 1.0000 | 0.9953 | 0.8770 | 0.8979 |
| 70 | 0.9142 | 0.9545 | 1.0000 | 0.9953 | 0.7271 | 0.7239 |
| 71 | 0.9974 | 0.9088 | 0.9953 | 0.9937 | 0.8856 | 0.9048 |
| 72 | 0.9898 | 0.9071 | 0.9953 | 0.9937 | 0.8609 | 0.8809 |
| 73 | 0.9908 | 0.8923 | 0.9953 | 0.9937 | 0.8555 | 0.8837 |
| 74 | 0.9811 | 0.9740 | 0.9953 | 0.9937 | 0.8479 | 0.8321 |
| 75 | 0.9877 | 0.9169 | 0.9953 | 0.9937 | 0.7210 | 0.7332 |
| 76 | 0.9811 | 0.9740 | 0.9953 | 0.9937 | 0.8454 | 0.8297 |
| 77 | 0.9805 | 0.9680 | 0.9926 | 0.9890 | 0.7502 | 0.7400 |
| 78 | 0.9894 | 0.8684 | 0.9926 | 0.9890 | 0.9428 | 0.9907 |
| 79 | 0.9767 | 0.9422 | 0.9926 | 0.9890 | 0.8484 | 0.8507 |
| 80 | 0.9877 | 0.9281 | 0.9926 | 0.9890 | 0.7418 | 0.7500 |
| 81 | 0.9917 | 0.9185 | 0.9896 | 0.9906 | 0.8629 | 0.8778 |
| 82 | 0.9967 | 0.8939 | 0.9896 | 0.9906 | 0.8175 | 0.8447 |
| 83 | 0.9965 | 0.9560 | 0.9896 | 0.9906 | 0.8094 | 0.8039 |
| 84 | 0.9857 | 0.9754 | 0.9896 | 0.9906 | 0.7838 | 0.7696 |
| 85 | 0.9937 | 0.9185 | 0.9736 | 0.9676 | 0.8207 | 0.8409 |
| 86 | 0.9859 | 0.9798 | 0.9736 | 0.9676 | 0.8078 | 0.7966 |
| 87 | 0.9956 | 0.9039 | 0.9736 | 0.9676 | 0.6722 | 0.6952 |
| 88 | 0.9897 | 0.8973 | 0.9736 | 0.9676 | 0.6704 | 0.6966 |
| 89 | 0.9943 | 0.9022 | 0.9759 | 0.9716 | 0.8905 | 0.9209 |
| 90 | 0.9906 | 0.9297 | 0.9759 | 0.9716 | 0.7109 | 0.7224 |
| 91 | 1.0000 | 0.9120 | 0.9759 | 0.9716 | 0.6500 | 0.6678 |
| 92 | 0.9915 | 0.9055 | 0.9759 | 0.9716 | 0.6187 | 0.6386 |
| 93 | 0.9491 | 0.9620 | 0.9948 | 0.9930 | 0.9000 | 0.8915 |
| 94 | 0.9683 | 0.9605 | 0.9948 | 0.9930 | 0.7617 | 0.7545 |
| 95 | 0.9055 | 0.9590 | 0.9948 | 0.9930 | 0.7516 | 0.7475 |
| 96 | 0.9142 | 0.9695 | 0.9948 | 0.9930 | 0.7512 | 0.7419 |
| 97 | 0.8992 | 0.9590 | 0.9962 | 0.9941 | 0.9598 | 0.9544 |
| 98 | 0.9762 | 0.9249 | 0.9962 | 0.9941 | 0.9478 | 0.9593 |
| 99 | 0.9317 | 0.9560 | 0.9962 | 0.9941 | 0.7780 | 0.7738 |
| 100 | 0.8433 | 0.9650 | 0.9962 | 0.9941 | 0.8951 | 0.8892 |

**TABLE 6.3 The Common Weights of Two Different Forms of Auto Sales Function**

| No. | Effect factor | Linear form | Exponential form |
|---|---|---|---|
| 1 | Price | 0.5943 | 0.0500 |
| 2 | Quality | 0.2068 | 0.6284 |
| 3 | SSI | 0.1089 | 0.1559 |
| 4 | ASI | 0.1989 | 0.2039 |
| 5 | Efficiency score | 0.0083 | 0.6386 |

Table 6.3. Under the linear hypotheses of the form of the sales function, price influences sales dramatically, and quality, ASI, and SSI subsequently. On the condition that the sales function is exponential, quality has a significant impact on sales; then ASI, SSI, and price are ranked in order of importance. The result is consistent with Figure 6.1.

With the assumption of exponential auto sales function, the specific form of sales function is obtained as the following equation.

$$S' = P'^{0.0500} \times Q'^{0.6284} \times SSI'^{0.1559} \times ASI'^{0.2039} \tag{6.17}$$

where $S'$, $P'$, $Q'$, $SSI'$, $ASI'$ respectively denote the values that are transformed and normalized using the above-mentioned method, while $S, P, Q, SSI, ASI$ represent the real sales, price, quality, SSI, and ASI in the auto industry, separately.

From the above analysis, on the one hand, price has a diminutive impact on sales and does not reflect the public product information in the Chinese automobile market. Moreover, the efficiency of the whole automobile market is low as the product efficiency scores of most vehicle types are not high. On the other hand, due to foreign manufacturers having advance technology and manufacturing competition, the Chinese government promulgates many special automotive industry policies to improve market efficiency and protect Chinese automotive manufacturers from losing market shares and international competitiveness. For example, all cars used for public affairs in the Chinese party or government organizations are Chinese self-owned brands. Especially with the worsening of the environment and traffic in China in recent years, more and more relevant regulations and control policies, such as policies restricting car purchasing, are proposed to alleviate this situation. Thus the Chinese automobile market is influenced heavily by the Chinese government rather than market mechanisms. In conclusion, EMH does not stand in the Chinese automobile market.

Using the efficiency scores of the 7th column in Table 6.2, the average efficiency of each auto brand is calculated and the results are showed in Table 6.4. Table 6.4 indicates that the average efficiency of every auto brand

**TABLE 6.4 The Average Efficiency of Each Auto Brand**

| No. | Brand | Manufacturer | Group | Efficiency scores |
|---|---|---|---|---|
| 1 | Hyundai | Beijing Hyundai | Beijing Auto | 0.8723 |
| 2 | BYD | BYD | BYD | 0.8461 |
| 3 | Nissan | Dongfeng Nissan | Dongfeng Group | 0.8786 |
| 4 | Kia | DYK | DYK | 0.8401 |
| 5 | Honda | Guangqi Honda | Guangqi | 0.8553 |
| 6 | JAC | JAC | JAC | 0.8528 |
| 7 | Chery | Chery Auto | Chery Auto | 0.8310 |
| 8 | VW | SVW | Saic Group | 0.9441 |
| 9 | Skoda | SVW | Saic Group | 0.8889 |
| 10 | Roewe | Saic Motor | Saic Group | 0.7921 |
| 11 | Buick | SGM | Saic Group | 0.9391 |
| 12 | Chevrolet | SGM | Saic Group | 0.9004 |
| 13 | Toyota | FAW Toyota | FAW | 0.8486 |
| 14 | Chana | Chana Auto | Chana Auto | 0.8441 |
| 15 | Great Wall | Great Wall | Great Wall | 0.8329 |
| 16 | LiFan | LiFan Auto | LiFan Auto | 0.8240 |
| 17 | Zotye | Zotye Auto | Zotye Auto | 0.7573 |
| 18 | Changhe | Changhe Auto | Changhe Auto | 0.7374 |
| 19 | Honda | Dongfeng Honda | Dongfeng Group | 0.7838 |
| 20 | Toyota | GTMC | GTMC | 0.8942 |

is not very high, which also illustrates that EMH is invalid in the Chinese automobile market.

Table 6.4 shows that some manufacturers belong to the same Auto group. For example, Dongfeng Nissan and Donfeng Honda are Dongfeng Groups's subsidiaries. Thus we regard No. 3 (Nissan) and No. 19 (Honda as Dongfeng Group's brands. SVW, Saic Motor, and SGM are all Saic Group's subsidiaries. We treat Nos. 8–12 as Saic Group's brands. It is interesting that No. 13 (Toyota) and No. 20 (Toyota) are different Auto groups' brands, since both FAW and GTMC owns the brand permit. Considering the distributions of subsidiaries of different Auto groups, we regard those brands that belong to the four automobile groups (FAW, Saic Motor, Dongfeng, Chana Auto) as the national brands that cover the whole country in this study, and others are local brands that only cover local areas. By using the results in Table 6.4, we obtain the average efficiency of national brands as 0.8689, while the average efficiency of local brands is 0.8288 and smaller than the former. Results indicate that the national brands are more efficient than local brands. This conclusion offers one possible explanation as to why the Chinese government encourages the four automobile groups to carry out the automotive enterprise's merger and reorganization across the country.

From the view of the property of enterprises, auto brands can be divided into two distinct categories: joint venture brand and self-owned brand. Beijing Hyundai, Dongfeng Nissan, DYK, Guangqi Honda, SVW, SGM, FAW Toyota, Dongfeng Honda, GTMC are joint venture brand, and the residual are self-owned brands. Similarly, the average efficiency of joint venture brands is 0.8769, while the average efficiency of self-owned brands is 0.8131; thus, we tentatively reach the conclusion that the efficiency of joint venture brands is higher than self-owned brands. This result may be attributed to the advanced technology and high quality possessed by joint venture brands.

From Figures 6.2 and 6.3, we can see that self-owned and local brands have lower average efficiency scores (the value is 0.8116), while the brands

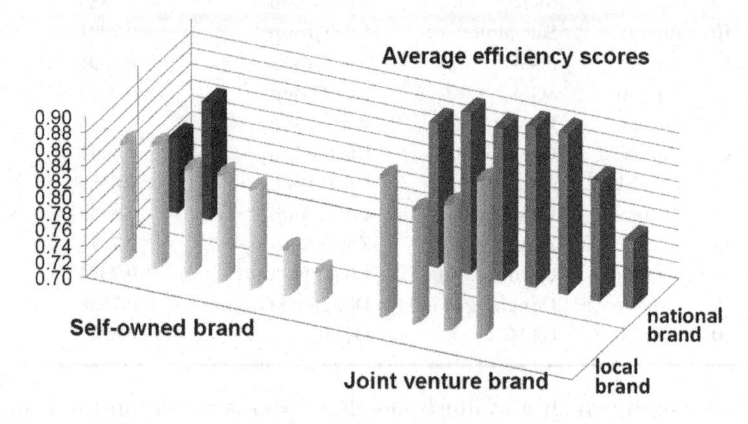


**Figure 6.2** Average efficiency scores of all auto brands.

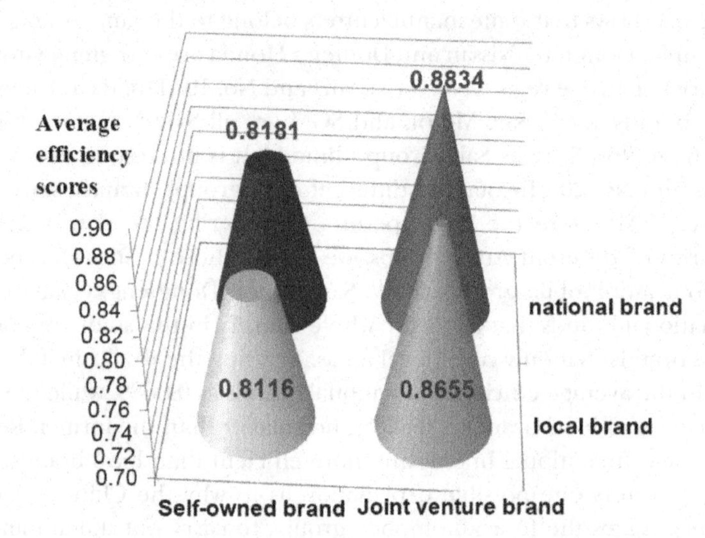


**Figure 6.3** Average efficiency scores of the brand classifications.

that are joint venture enterprises and possess national scale have higher average efficiency scores (the value is 0.8834). It demonstrates that the brands that own large automobile brand scales, advanced technology, and high auto quality are more efficient in the Chinese auto market.

## CONCLUSION

Combining the efficiency market hypothesis with DEA-based technique, this chapter studies the efficiency of the Chinese automobile market and describes the relationship between auto price, quality, SSI, ASI, and auto sales on the basis of correlatively available auto data for a sample of 100 vehicle types in 2012. Using a real dataset, it has demonstrated that the relationship among these variables is adequately elaborated by the exponential form and that quality is the most major influencing factor, followed by ASI. The result is helpful to provide some managerial implications for decision makers and sales managers in the automobile industry. They should consider the quality of vehicles as the first thing for improving auto sales. Moreover, as the efficiency score of each vehicle type solving by Model (15) is low, it is safe to say that EMH does not coincide well with the Chinese auto market. In the view of the scale of brand, national brands are more efficient than local brands, and the average efficiency of self-owned brands is lower than that of the joint venture brands from the perspective of the property of enterprises additionally. Moreover, national and joint venture brands have a higher market efficiency compared with other brands. The results imply that the Chinese government should encourage and support automotive enterprises' merger and reorganization across the country and technology research and development if it wants to promote the efficiency of the Chinese auto market.

Finally, we discuss the limitation of the work and potential future directions. Due to automobiles being different from one another in some characteristics, every vehicle type has several cars that have little performance discrepancy. Based on this, the auto price and quality used in this chapter is the average price of each vehicle type. Hence, results may be changed since there exist certain inaccuracies on data collection and calculation. In addition, it should be noticed that this study has examined only the influence factors from the consumers' perspective and ignored the efforts of manufacturers, such as promoting activities or advertising. Hence, one line of analysis for further study is to consider or increase other independent variables, such as marketing expense, advertising, and brand effect.

**APPENDIX 1 The Data Set of 100 Vehicle Types in China**

| No | Vehicle type | Brand | Manufacturer | Price | Quality | SSI | ASI | Sales |
|---|---|---|---|---|---|---|---|---|
| 1 | i30 | Hyundai | Beijing Hyundai | 19.5327 | 4.15 | 839 | 859 | 3200 |
| 2 | Mdavante | Hyundai | Beijing Hyundai | 14.3208 | 4.04 | 839 | 859 | 80460 |
| 3 | Moinca | Hyundai | Beijing Hyundai | 33.0139 | 4.42 | 839 | 859 | 16892 |
| 4 | Verna | Hyundai | Beijing Hyundai | 20.2455 | 4.41 | 839 | 859 | 201746 |
| 5 | Sonata | Hyundai | Beijing Hyundai | 14.9386 | 4.08 | 839 | 859 | 100454 |
| 6 | Elantra | Hyundai | Beijing Hyundai | 20.1663 | 4.02 | 839 | 859 | 64363 |
| 7 | NewElantra | Hyundai | Beijing Hyundai | 19.1366 | 4.10 | 839 | 859 | 213974 |
| 8 | F0 | BYD | BYD | 17.7980 | 4.08 | 765 | 791 | 46956 |
| 9 | F3 | BYD | BYD | 11.3109 | 4.12 | 765 | 791 | 123424 |
| 10 | G3 | BYD | BYD | 11.2317 | 4.53 | 765 | 791 | 31964 |
| 11 | G3R | BYD | BYD | 13.1327 | 4.35 | 765 | 791 | 3222 |
| 12 | G6 | BYD | BYD | 10.5980 | 4.27 | 765 | 791 | 41075 |
| 13 | L3 | BYD | BYD | 9.8059 | 3.62 | 765 | 791 | 77867 |
| 14 | Speed sharp | BYD | BYD | 15.8099 | 4.18 | 765 | 791 | 38664 |
| 15 | Livina | Nissan | Dongfeng Nissan | 20.9188 | 4.36 | 836 | 857 | 48225 |
| 16 | March | Nissan | Dongfeng Nissan | 15.4931 | 4.31 | 836 | 857 | 6720 |
| 17 | Tidda | Nissan | Dongfeng Nissan | 21.2911 | 4.07 | 836 | 857 | 144215 |
| 18 | Teana | Nissan | Dongfeng Nissan | 43.5327 | 4.41 | 836 | 857 | 90072 |
| 19 | Sylphy | Nissan | Dongfeng Nissan | 16.0792 | 4.47 | 836 | 857 | 133823 |
| 20 | Sunny | Nissan | Dongfeng Nissan | 12.8950 | 4.16 | 836 | 857 | 137820 |
| 21 | K3 | Kia | DYK | 14.4792 | 4.33 | 752 | 779 | 26849 |
| 22 | K5 | Kia | DYK | 17.5525 | 4.28 | 752 | 779 | 52745 |
| 23 | RIO | Kia | DYK | 32.4436 | 4.22 | 752 | 779 | 9184 |
| 24 | Forte | Kia | DYK | 17.6317 | 4.27 | 752 | 779 | 80989 |
| 25 | K2 | Kia | DYK | 14.4792 | 4.28 | 752 | 779 | 124941 |

*(continued)*

**APPENDIX 1 The Data Set of 100 Vehicle Types in China (continued)**

| No | Vehicle type | Brand | Manufacturer | Price | Quality | SSI | ASI | Sales |
|---|---|---|---|---|---|---|---|---|
| 26 | Cerato | Kia | DYK | 18.3446 | 4.02 | 752 | 779 | 35461 |
| 27 | Soul | Kia | DYK | 13.8297 | 3.78 | 752 | 779 | 10826 |
| 28 | Fit | Honda | Guangqi Honda | 38.7010 | 4.04 | 787 | 785 | 35920 |
| 29 | Crosstour | Honda | Guangqi Honda | 15.8099 | 4.27 | 787 | 785 | 33937 |
| 30 | Crider | Honda | Guangqi Honda | 54.5426 | 4.55 | 787 | 785 | 24576 |
| 31 | City | Honda | Guangqi Honda | 20.3248 | 4.48 | 787 | 785 | 96913 |
| 32 | Accord | Honda | Guangqi Honda | 20.9584 | 4.51 | 787 | 785 | 104114 |
| 33 | Amiable | JAC | JAC | 8.3644 | 3.66 | 729 | 806 | 50088 |
| 34 | Lucky bell | JAC | JAC | 6.9386 | 4.41 | 729 | 806 | 42345 |
| 35 | RS | JAC | JAC | 19.1366 | 4.23 | 729 | 806 | 36037 |
| 36 | Yueyue | JAC | JAC | 10.4238 | 3.63 | 729 | 806 | 13783 |
| 37 | A1 | Chery | Chery Auto | 7.0337 | 4.26 | 818 | 799 | 18282 |
| 38 | A3 | Chery | Chery Auto | 11.4535 | 4.11 | 818 | 799 | 17901 |
| 39 | E5 | Chery | Chery Auto | 8.5228 | 4.16 | 818 | 799 | 64241 |
| 40 | QQ | Chery | Chery Auto | 9.7901 | 4.40 | 818 | 799 | 143974 |
| 41 | Fungwan2 | Chery | Chery Auto | 7.5089 | 3.94 | 818 | 799 | 23836 |
| 42 | Cown1 | Chery | Chery Auto | 13.9881 | 4.14 | 818 | 799 | 15536 |
| 43 | Cown2 | Chery | Chery Auto | 14.0673 | 4.01 | 818 | 799 | 33163 |
| 44 | Cown3 | Chery | Chery Auto | 18.3446 | 4.06 | 818 | 799 | 19091 |
| 45 | Lavida | VW | Shanghai-VW | 21.7584 | 4.37 | 831 | 863 | 246687 |
| 46 | Passat | VW | Shanghai-VW | 40.2059 | 4.46 | 831 | 863 | 233321 |
| 47 | Santana | VW | Shanghai-VW | 26.3683 | 4.21 | 831 | 863 | 104304 |
| 48 | Santana Vista | VW | Shanghai-VW | 19.3901 | 4.11 | 831 | 863 | 98484 |
| 49 | Polo | VW | Shanghai-VW | 16.5307 | 4.27 | 831 | 863 | 119173 |
| 50 | Superb | Skoda | Shanghai-VW | 15.4139 | 4.11 | 799 | 812 | 44007 |

*(continued)*

**APPENDIX 1 The Data Set of 100 Vehicle Types in China (continued)**

| No | Vehicle type | Brand | Manufacturer | Price | Quality | SSI | ASI | Sales |
|---|---|---|---|---|---|---|---|---|
| 51 | Fabia | Skoda | Shanghai-VW | 15.4218 | 4.07 | 799 | 812 | 48380 |
| 52 | Octavia | Skoda | Shanghai-VW | 32.9347 | 4.45 | 799 | 812 | 137616 |
| 53 | 350 | Roewe | Saic Motor | 17.1406 | 4.25 | 825 | 839 | 74384 |
| 54 | 550 | Roewe | Saic Motor | 40.3010 | 4.61 | 825 | 839 | 38581 |
| 55 | 950 | Roewe | Saic Motor | 35.2950 | 4.41 | 825 | 839 | 4905 |
| 56 | Regal | Buick | SGM | 16.8554 | 4.32 | 815 | 852 | 85440 |
| 57 | LaCrosse | Buick | SGM | 10.7802 | 3.98 | 815 | 852 | 86101 |
| 58 | Excelle | Buick | SGM | 21.2911 | 4.29 | 815 | 852 | 277071 |
| 59 | ExcelleXT | Buick | SGM | 25.1248 | 4.24 | 815 | 852 | 127575 |
| 60 | Epica | Chevrolet | SGM | 47.2713 | 4.43 | 824 | 830 | 18580 |
| 61 | Cruze | Chevrolet | SGM | 37.9248 | 4.24 | 824 | 830 | 232592 |
| 62 | Sall | Chevrolet | SGM | 15.5723 | 4.23 | 824 | 830 | 279740 |
| 63 | Aveo | Chevrolet | SGM | 31.3505 | 4.45 | 824 | 830 | 56627 |
| 64 | Malibu | Chevrolet | SGM | 24.2218 | 4.03 | 824 | 830 | 51926 |
| 65 | Cololla | Toyota | FAW Toyota | 16.2059 | 4.24 | 840 | 836 | 124531 |
| 66 | Crown | Toyota | FAW Toyota | 66.5030 | 4.52 | 840 | 836 | 25456 |
| 67 | Cololla | Toyota | FAW Toyota | 22.5426 | 4.28 | 840 | 836 | 151887 |
| 68 | Prius | Toyota | FAW Toyota | 97.4099 | 4.50 | 840 | 836 | 18706 |
| 69 | Reiz | Toyota | FAW Toyota | 41.5525 | 4.03 | 840 | 836 | 59846 |
| 70 | RAV4 | Toyota | FAW Toyota | 36.1663 | 4.30 | 840 | 836 | 9130 |
| 71 | RushingMini | Chana | Chana Auto | 6.7960 | 4.01 | 814 | 827 | 66667 |
| 72 | Alsvin | Chana | Chana Auto | 9.9644 | 4.00 | 814 | 827 | 48878 |
| 73 | AlsvinV3 | Chana | Chana Auto | 9.5683 | 3.91 | 814 | 827 | 45656 |
| 74 | CX20 | Chana | Chana Auto | 13.4495 | 4.43 | 814 | 827 | 41511 |
| 75 | CX30 | Chana | Chana Auto | 10.8198 | 4.06 | 814 | 827 | 8458 |

*(continued)*

**APPENDIX 1 The Data Set of 100 Vehicle Types in China (continued)**

| No | Vehicle type | Brand | Manufacturer | Price | Quality | SSI | ASI | Sales |
|---|---|---|---|---|---|---|---|---|
| 76 | Eado | Chana | Chana Auto | 13.4495 | 4.43 | 814 | 827 | 40239 |
| 77 | C20R | Great Wall | Great Wall | 13.6871 | 4.39 | 799 | 801 | 12189 |
| 78 | C30 | Great Wall | Great Wall | 10.1228 | 3.77 | 799 | 801 | 136502 |
| 79 | C50 | Great Wall | Great Wall | 15.1762 | 4.22 | 799 | 801 | 41802 |
| 80 | Ling Ao | Great Wall | Great Wall | 10.8198 | 4.13 | 799 | 801 | 10978 |
| 81 | 620 | LiFan | LiFan Auto | 9.1644 | 4.07 | 783 | 810 | 50127 |
| 82 | 320 | LiFan | LiFan Auto | 7.0812 | 3.92 | 783 | 810 | 28365 |
| 83 | 520 | LiFan | LiFan Auto | 7.1921 | 4.31 | 783 | 810 | 25625 |
| 84 | 720 | LiFan | LiFan Auto | 11.6119 | 4.44 | 783 | 810 | 18591 |
| 85 | T200 | Zotye | Zotye Auto | 8.3327 | 4.07 | 703 | 693 | 29535 |
| 86 | Z300 | Zotye | Zotye Auto | 11.5644 | 4.47 | 703 | 693 | 25126 |
| 87 | Z200 | Zotye | Zotye Auto | 7.5723 | 3.98 | 703 | 693 | 4587 |
| 88 | Z200HB | Zotye | Zotye Auto | 9.9960 | 3.94 | 703 | 693 | 4483 |
| 89 | Wagon | Changhe | Changhe Auto | 8.0871 | 3.97 | 714 | 712 | 70886 |
| 90 | Liana | Changhe | Changhe Auto | 9.6317 | 4.14 | 714 | 712 | 7451 |
| 91 | Adel | Changhe | Changhe Auto | 5.6871 | 4.03 | 714 | 712 | 3470 |
| 92 | Splash | Changhe | Changhe Auto | 9.2515 | 3.99 | 714 | 712 | 2344 |
| 93 | Civic | Honda | Dongfeng-Honda | 25.1564 | 4.35 | 811 | 823 | 79763 |
| 94 | Ciimo | Honda | Dongfeng-Honda | 18.3446 | 4.34 | 811 | 823 | 14089 |
| 95 | Spirior | Honda | Dongfeng-Honda | 38.6218 | 4.33 | 811 | 823 | 12417 |
| 96 | CR-V | Honda | Dongfeng-Honda | 36.1663 | 4.40 | 811 | 823 | 12341 |
| 97 | Camry | Toyota | GTMC | 40.3644 | 4.33 | 819 | 829 | 169036 |
| 98 | Yaris | Toyota | GTMC | 15.3505 | 4.11 | 819 | 829 | 145402 |
| 99 | Verso | Toyota | GTMC | 30.8594 | 4.31 | 819 | 829 | 17286 |
| 100 | Highlander | Toyota | GTMC | 53.7505 | 4.37 | 819 | 829 | 75059 |

## ACKNOWLEDGEMENTS

The financial supports from National Natural Science Foundation of China (Grant Nos. 71322101, 71271195 and 71110107024) are acknowledged.

## REFERENCES

Banker, R. D., Das, S., & Datar, S. (1989). Analysis of cost variances for management control in hospitals. *Research in Governmental and Nonprofit Accounting, 5*, 268–291.

Case, K. E., & Shiller, R. J. (1989). The efficiency of the market for single-family homes. *The American Economic Review, 79*(1), 125–137.

Charnes, A., Clarke, C., Cooper, W. W., & Golany, B. (1985). A development study of DEA in measuring the effect of maintenance units in the US air force. *Annals of Operations Research, 2*(1), 95–112.

Charnes, A., Cooper, W. W., & Rhodes, E. (1978). Measuring the efficiency of decision making units. *European Journal of Operational Research, 2*(6), 429–444.

China Association of Automobile Manufacturers. (2012). *The Annual Report of China Association of Automobile Manufacturers 2012.* Beijing, China: Author.

Choi, H., & Oh, I. (2010). Analysis of product efficiency of hybrid vehicles and promotion policies. *Energy Policy, 38*(5), 2262–2271.

Cook, W. D., & Green, R. H. (2004). Multi component efficiency measurement and core business identification in multi plant firms: A DEA model. *European Journal of Operational Research, 157*(3), 540–551.

Emrouznejad, A., Parker, B. R., & Tavares, G. (2008). Evaluation of research in efficiency and productivity: A survey and analysis of the first 30 years of scholarly literature in DEA. *Socio-Economic Planning Sciences, 42*(3), 151–157.

Ertay, T., Ruan, D., & Tuzkaya, U. R. (2006). Integrating data envelopment analysis and analytic hierarchy for the facility layout design in manufacturing systems. *Information Sciences, 176*(3), 237–262.

Fama, E. F. (1970). Efficient capital markets: a review of theory and empirical work. *The Journal of Finance, 25*(2), 383–417.

Guntermann, K. L., & Norrbin, S. C. (1991). Empirical tests of real estate market efficiency. *The Journal of Real Estate Finance and Economics, 4*(3), 297–313.

Jensen, M. C. (1978). Some anomalous evidence regarding market efficiency. *Journal of Financial Economics, 6*(2), 95–101.

Oh, I., Lee, J. D., Hwang, S., & Heshmati, A. (2010). Analysis of product efficiency in the Korean automobile market from a consumer's perspective. *Empirical Economics, 38*(1), 119–137.

Palan, S. (2004). *The efficient market hypothesis and its validity in today's markets.* Santa Cruz, CA: Grin Verlag.

Ray, S. C. (2004). *Data envelopment analysis: theory and techniques for economics and operations research.* Cambridge, UK: Cambridge University Press.

Saranga, H. (2009). The Indian auto component industry: Estimation of operational efficiency and its determinants using DEA. *European Journal of Operational Research, 196*, 707–718.

Seiford, L. M., & Zhu, J. (1999). Profitability and marketability of the top 55 US commercial banks. *Management Science, 45*(9), 1270–1288.

State Council of China. (1986). *The seventy five-year plan for national economic and social development of the People's Republic China.* Beijing, China: National Development and Reform Commission.

Tian, L. (2007). Does government intervention help the Chinese automobile industry? A comparison with the Chinese computer industry. *Economic Systems, 31*, 364–374.

Wang, J., Liu, S., Wang, Y., & Xie, N. (2008). Evaluation of customer satisfaction in automobile after-sales service based on grey incidence analysis. In *IEEE International Conference Systems, Man and Cybernetics* (pp. 2386–2389). New York, NY: IEEE.

Yang, C., Yang, F., Xia, Q. & Ang S. (2012). What makes sales in Chinese shampoo industry? A DEA study based on efficient market hypothesis. *Asia Pacific Journal of Marketing and Logistic, 24*(4), 678–689.

CHAPTER 7

# A CLUSTERING ANALYSIS OF FIVE-STAR MORNING STAR RULED MODERATE ASSET ALLOCATION FUNDS

**Kenneth D. Lawrence**
***New Jersey Institute of Technology***

**Gary Kleinman**
***Montclair State University***

**Sheila M. Lawrence**
***Rutgers University***

## ABSTRACT

The purpose of this chapter is to determine whether a seemingly uniform group of 16 mutual funds, denoted by Morningstar as five-star, moderate asset allocation mutual funds, is actually heterogeneous in composition. Fifty-three financial variables were collected on each fund. These variables include measures of performance, risk, and equity composition and bond composition. The data were analyzed using various statistically based clustering measures and distance functions. The results revealed three clusterings within

*Contemporary Perspectives in Data Mining, Volume 2*, pages 123–135

the Morningstar classification. The implications of this clustering for mutual fund selection tools are described.

## INTRODUCTION

It is the aim of this research to analyze the differences among a group of 16 Morningstar five-star rated moderate allocation funds. Each of these funds has a set of 53 financial variables. These variables include measures of performance, risk, and equity composition and bond composition.

Using various statistical clustering procedures and distance functions, the 16 five-star funds are to be grouped into like clusters of funds. These like groups will consist of funds whose behavioral and performance characteristics are quite similar in nature. Accordingly, funds within the same grouping are relatively the same as, and therefore somewhat duplicative of, other funds within the same grouping.

## A MORNINGSTAR ASSET ALLOCATION FUND

The Morningstar system for categorizing and rating mutual funds has been a significant part of the investment landscape for many years, with its judgments about risk greatly influencing investor behavior (e.g., Sharpe, 1998). The so-called Morningstar rating for mutual funds assigns from one to five stars to funds based on Morningstar's evaluation of the fund characteristics of return and risk. Return and risk, though, Morningstar maintains, are best assessed when the return and risk characteristics of specific funds are compared to the return and risk characteristics of funds whose managers employ the same investment strategy. Morningstar assigns funds to one of six categories. This categorization process lumps funds with similar investment strategies into the same category. Within those categories, an investor can compare funds based on so-called risk adjusted and cost-adjusted ratings. One such category is moderate asset allocation funds, the category of key interest in this study.

Given the importance of risk adjusted ratings in the star-assignment process, it is important to understand *how* Morningstar calculates risk. Morningstar (2010) states that its rating is based on "expected utility theory." Under expected utility theory,

> [I]nvestors are a) more concerned about a possible poor outcome than an unexpectedly good outcome and b) more willing to give up some portion of their expected return in exchange for greater certainty of return. The rating accounts for all variations in a fund's monthly performance, with more emphasis on downward variations. It rewards consistent performance and re-

duces the risk of strong short-term performance masking the inherent risk of a fund. (p. 1)

Further, the methodology used takes into account sales charges, redemption fees, and loads. The age of the fund influences the weights assigned to the annual performance of the fund. Morningstar (2010), for example, states that a fund that is at least three years old but less than five years old would have 100% of its weighting based on the first three years of its existence.

However, if the fund was at least five years old but less than ten years, then 60% of the firm's weighting would be based on the five-year rating, and 40% on the three-year rating. Should the fund be at least 10 years old, the weight distribution scheme would assign 50% of the rating to the ten year performance, 30% to the five-year performance, and 20% to the three-year performance. Note, though, that these ratings—which are retrospective in nature—are separate from the Morningstar analyst ratings of funds. The latter ratings have a strong prospective element.

Our key interest, though, is in the categorization of funds by Morningstar. Mutual funds generally are broken down into three classes, based on the asset type predominantly held within it. The three broad classes are stocks, bonds and cash (see Thune, 2013). Stock funds, the object of interest here, are further broken down based on the market capitalization of stocks the funds invest in, with market capitalization calculated as the average price of each share times the number of shares outstanding. The three capitalization categories that Morningstar uses are (1) large capitalization, (2) mid-cap capitalization, and (3) small capitalization. Large capitalization stock funds are funds with greater than $11 billion in assets, according to Morningstar, cited in Thune (n.d.). Midcap stock funds invest in firms with intermediate-sized capitalization. For Morningstar, that suggests capitalization between $2.5 billion and $11 billion. Finally, small capitalization stock funds have asset sizes from $750 million up to $2.5 billion dollars.

These categories can be broken down still further based on the fund management's investment objective. These objectives can be growth-oriented—that is, the objective is aimed toward acquiring stocks expected to grow faster than the market average; value objective, which means the objective is to purchase stocks that the manager believes are undervalued by the market; and, finally, blended stock funds, which invest in a mixture of growth and value stocks. Funds can be further categorized by whether they are large capitalization AND value stocks, and so on.

Our interest here is in five-star moderate allocation funds. Moderate allocation funds are defined by Carlson (2013), writing in a Morningstar newsletter, as funds that "have equity stakes of 50% to 70%." Specifically, we intend to use cluster analysis on our data and compare the clusters with the intended uniformity of Morningstar's own classification of funds as five-star, moderate allocation funds. In effect, we are investigating the diversity that exists within a uniform asset classification. Understanding the diversity of assets within a Morningstar categorization (5-star, moderate allocation) is important because Morningstar ratings have been found to be consequential. For example, Morningstar ratings have been found to have enough credibility to result in management removal. Barron and Ni (2013) found that Morningstar ratings had a significant impact on mutual fund manager turnover in the predicted direction. This research built on earlier research that found that investors themselves respond to Morningstar ratings; that is, funds rated better by Morningstar had greater investor interest than funds that received relatively poorer ratings. We found only three prior studies that sought to compare statistically generated groups of funds with Morningstar's own clustering of funds.

Otranto (2008) studied time series data, using model-based methods. In these methods, similarity between models characterizing the data is found related to similarity of time series modeled. He notes that, when dealing with time series data, investors face huge investment universes and therefore prefer to face "groups of series with similar characteristics" (p. 4685). Otranto further notes the utility of classifying time series in homogeneous clusters when the characteristic of interest is similarity of volatility structures.

In Otranto's (2008) paper, however, he argues the need for a clustering procedure based on the "squared disturbance of the returns of a financial time series as the volatility of the series" (p. 4686). In conducting his analysis using Garch representation, he found he could separate the characteristic of interest, volatility, into two parts: a time-varying part and a constant part. Whereas the constant part reflects minimum, Otranto notes that the key value of interest is unconditional volatility. He approaches his task using a beginning series. He then applies an "agglomerative algorithm" to the results, generating three clustering levels. With Otranto's approach, the algorithm, not the user, determines the number of clusters or level of deepness detected. Deepness, in Otranto's definition, implies many clusters, each with small groups of resident members.

Whereas Otranto examined subsets of volatility using time series data, however, we seek to understand the number and divergence of clusters existing within a set of mutual funds that Morningstar has characterized as homogeneous with respect to investment strategy (moderate risk allocation choice) and performance (five star, the top level of performance.) While Morningstar takes risk and return, as well as fees, into account in assigning funds to star categories, and further takes asset allocation aggressiveness into account in assigning funds to conservative versus moderate versus aggressive categories, we seek to further the field's understanding of potential diversity of characteristics within a single cross-joined cell of Morningstar firms, the 5-star moderate asset allocation firms.

Otranto (2008) took one relatively limited approach to the use of cluster analysis to gain better understanding of investment vehicles using time series, return, and volatility data. Pattarin, Paterlini, and Minerva (2004) argue for a broader use of the power of cluster analysis to detect underlying patterns, or groupings, of mutual funds. Pattarin et al. note that the classificatory powers that cluster analysis provides can be useful when contexts of interest are beset by high degrees of complexity. They argue that the set of statistical techniques they use in their study yield a better understanding of the objects of interest, financial products, at relatively low cost. Their argument for the need for their set of techniques is consistent with the motivation for our study. "Institutional classification schemes," Pattarin et al. (2004) state, "when available, do not always provide consistent and representative peer groups of funds" (p. 353). Similarly, we argue that the Morningstar classification scheme obscures diversity resident in the firms classified. Just as Pattarin et al. argue that their return-based classification scheme for analyzing mutual fund styles based on past returns could provide more reliable information to investors, we argue that understanding subgroups within a particular cell in the Morningstar star/aggressiveness classification can provide more reliable information to investors as well. To the extent that investor markets prefer granularity in the information they consume in making investment choices, our study provides a means of generating this granularity.

Pattarin et al.'s (2004) method is composed of three steps. The first step uses principal component analysis to discern underlying dimensions in the data. The second step employs a robust evolutionary clustering methodology to determine clusters within the sample of Italian mutual funds. The evolutionary clustering methodology employed is called the Genetic Algorithm for Medoid Evolution. With genetic algorithms, iterative search processes continue until a termination condition is achieved. Third, the authors employ Sharpe's (1992) constrained regression model to analyze mutual fund investment style.

Investment styles range from conservative to very aggressive. In the Pattarin et al. study, as in ours, style (apart from active versus passive management) is determined by the investment manager's choice to invest heavily in more risky versus less risky return-generating assets. Like Otranto, Pattarin et al. examine time series of returns. Unlike our study, however, Pattarin et al. apply their methodology to several Assegestioni (the classifying body in Italy) categories and look to see whether there is significant overlap between the classifications the Pattarin et al. study assigns the Italian mutual funds to, and the Assegestioni's classification of the same funds. The authors found substantial overlap between Assegestioni's classification of the funds and that of Pattarin et al. This finding is relevant to the investor universe because the Pattarin et al. study employed more variables in its classification analysis than did Assegestioni. This suggests that the Pattarin et al. classifications were more robust and possessed more credibility than did the Assegestioni classification. Similarly, our study bears within itself the potential to support, or detract from, the credibility of the Morningstar classification schema.

The Pattarin et al. (2004) study examined the impact of time series returns and market indices (e.g., What percentage of equity-held assets were invested in Italy? Europe? The U.S.? The Pacific? Emerging markets? Were they held in Cash or in government bonds?) on the classification accuracy of Assegestioni's classification when the latter were compared with Pattarin et al.'s classification outcomes. Perez-Gladish, Rodriguez, M'Zali, and Lang (2013) studied whether there were differences in returns between firms that engaged in so-called socially responsible investing and firms that relied on financial criteria in making their investment decisions.

Socially responsible investing is defined by Perez-Gladish et al. (2013, p. 109) as "an investment process that integrates not only financial but also social, environmental, and ethical considerations into investment decision making." They note the continued growth of socially responsible investment in recent years. This, however, poses a problem for the subset of investors who seek to make socially responsible investments. The question that confronts them is: Who is to make the determination that an agency does make socially responsible investment decisions, and what are the criteria used? While investors could undertake to make the myriad investment analysis and classification decisions themselves, time is short and the number of classifications to be made is great. Instead, as Perez-Gladish et al. (2013) point out, investors need a tool that performs a classification function similar to what Morningstar performs in its classification of firms with respect to investing aggressiveness and performance given the investing aggressiveness classification.

The term *socially responsible investing* covers a wide range of potential behaviors. Perez-Gladish et al. (2013) limit their investigation to the environmental responsibility domain. The authors utilized an output-oriented radial data envelopment analysis (DEA) model with variable returns to scale. DEA is a methodology that permits evaluation of the comparative efficiency of decision-making units. The greater output achieved given a fixed input, the greater was the decision-making unit's efficiency. It is noted that studies of this sort use expected or excess return as the outcome or output of the mutual funds. Inputs include risk, cost,and measures of management processes related to socially responsible investing. We include risk, cost, and asset allocation measures in our evaluation of the uniformity of the Morningstar classification/performance cell under consideration here. Perez-Gadlish et al. concluded that there was no difference in relative performance outcomes of socially responsible mutual funds and other mutual funds that did not have socially responsible investments as a goal, as opposed to traditional financial performance. The Perez-Gadlish et al. (2013) study demonstrated the ability of advanced statistical techniques to answer the important question of whether socially responsible investing leads to poorer returns than do financial performance-related investing decisions.

## RESEARCH DATA

Data were collected from Morningstar. The data collected included all firms that received a five-star Morningstar rating that simultaneously were categorized as moderate allocation funds. Table 7.1, Panel A shows the list of funds that are both five-star funds and are categorized by Morningstar as "moderate allocation" funds. The table lists fund ID, type, and fund name. Table 7.1, Panel B describes the investment objectives of each mutual fund.

In order to determine whether meaningful subgroups of firms existed within the overall Morningstar five-star moderate asset group, we collected data on 53 variables. The variable list is shown in Table 7.2.

**TABLE 7.1 Panel A: List Of Funds That Are Both 5-Star and Moderate Allocation Funds**

| No. | Fund ID | Type | Name |
|---|---|---|---|
| 1 | RLBEX | Blend | American Funds America Funds Class |
| 2 | BUFBX | Value | Buffalo Flexible Income |
| 3 | CBAIZ | Blend | Columbia Balanaced Fund |
| 4 | HEFBX | Blend | Hennessy Equity and Income Fund Invester |
| 5 | FPACX | Blend | Crescent Fund |
| 6 | JABAX | Growth | Janus Balanced Fund |
| 7 | MTOIX | Values | Main Stay Income Builder Fund |
| 8 | PVSAX | Blend | Putnam Capital Spectrum |
| 9 | RNCOX | Blend | Core Opportunity Fund |
| 10 | FOBAX | Growth | Tributary Balanced Fund Instutional |
| 11 | VILLX | Growth | Villere Balanced Fund Investor |
| 12 | VWELX | Value | Vanguard Wellington Fund Investor |
| 13 | PRWCX | Blend | T. Rowe Price Capital Appreciation |
| 14 | MAPOX | Value | Marris & Power Balanced Fund Investor |
| 15 | WBALX | Growth | Weitz Balanced Fund |
| 16 | ACEIX | Value | Invesco Equity and Income Fund |

*(continued)*

**TABLE 7.1 Panel B: List of Objectives of Selected Funds (continued)**

| No. | Fund ID | Objectives From Plan Perspectus |
|---|---|---|
| 1 | RLBEX | Investment conservation of capital; current income; long-term capital gains and income |
| 2 | BUFBX | Long-term growth of capital |
| 3 | CBAIZ | Long-term growth of capital and current income |
| 4 | HBFBX | High total return |
| 5 | FPACX | Total return with reasonable risk through combination of income and capital appreciation |
| 6 | JABAX | Long-term capital growth with preservation of capital and current income |
| 7 | MTOIX | Current income consistent with growth of future capital and income |
| 8 | PVSAX | Total return |
| 9 | RNCOX | Long-term capital appreciation and income |
| 10 | FOBAX | Maximize total return consistent with generation of current income, preservation of capital, and reduced price volatility |
| 11 | VILLX | Long-term capital growth consistent with preservation of capital and balance of current income |
| 12 | VWELX | Provide long-term capital appreciation and reasonable current income |
| 13 | PRWCX | Long-term capital appreciation |
| 14 | MAPOX | Regular current income, potential capital appreciation at a moderate level |
| 15 | WBALX | Regular current income, capital preservation and long-term capital appreciation |
| 16 | ACEIX | Seek highest possible income consistent with safety of principal. Long-term growth of capital is secondary objective. |

**TABLE 7.2 Variables Used in Determining the Existence of Subgroups of Asset Allocation Funds from Fund Websites**

| No. | List of Variables |
|---|---|
| 1 | Net expense ratio |
| 2 | Total holdings |
| 3 | Portfolio turnover |
| 4 | Historic return |
| 5 | Historic risk |
| 6 | Asset allocation equity |
| 7 | Asset allocation bond |
| 8 | Asset allocation cash |
| 9 | Annualized return 1 yr. |
| 10 | Annualized return 3 yr. |
| 11 | Annualized return 5 yr. |
| 12 | Annualized return 10 yr. |
| 13 | Total assets |

*(continued)*

**TABLE 7.2 Variables Used in Determining the Existence of Subgroups of Asset Allocation Funds from Fund Websites (continued)**

| No. | List of Variables |
|---|---|
| 14 | U.S. assets |
| 15 | Market capitalization as a % of long equity assets—giant |
| 16 | Market capitalization as a % of long equity assets—large |
| 17 | Value & growth measure: price/prospective earnings |
| 18 | Value & growth measure: price/book |
| 19 | Value & growth measure: price/sales |
| 20 | Value & growth measure: price/cash flow |
| 21 | Value & growth measure: dividend yield |
| 22 | Value & growth measure: long term earning |
| 23 | Value & growth measure: historical earnings |
| 24 | Value & growth measure: sales growth |
| 25 | Value & growth measure: cash flow growth |
| 26 | Value & growth measure: book value & growth |
| 27 | Fixed income: corporate |
| 28 | Fixed income: securitized |
| 29 | Fixed income: government |
| 30 | Fixed income: cash and equivalence |
| 31 | Fixed income: municipal |
| 32 | Fixed income: derivative |
| 33 | Credit rating: AAA |
| 34 | Credit rating: A–AA |
| 35 | Maturity fixed income: 1–3 years |
| 36 | Maturity fixed income: 3–7 years |
| 37 | Equity sector: consumer discretionary |
| 38 | Equity sector: information technology |
| 39 | Equity sector: healthcare |
| 40 | Equity sector: financials |
| 41 | Equity sector: industrials |
| 42 | Equity sector: energy |
| 43 | Equity sector: consumer staples |
| 44 | Equity sector: materials |
| 45 | Equity sector: telecommunication services |
| 46 | Equity sector: utilities |
| 47 | Volatility measure 3 yr: standard deviation |
| 48 | Volatility measure 3 yr: return |
| 49 | Volatility measure 3 yr: sharpe ratio |
| 50 | MPT Statistics 3 yr R-square |
| 51 | MPT Statistics 3 yr beta |
| 52 | MPT Statistics 3 yr alpha |
| 53 | MPT statistics 3 yr treynor |

## CLUSTERING ANALYSIS METHODS

Clustering analysis methods were applied to the observations in order to determine whether subgroups of mutual funds existed within the conjoint Morningstar classification of fund deemed five-star and moderate asset allocation funds. Clustering analysis is a multivariate technique whose primary purpose is to group objects based on the characteristics they possess. The concept of similarity is fundamental to cluster analysis. Interobjective similarity is an empirical measure of resemblance between objects, which is measured by distance measures. The distance measures require metric data. These distance measures represent similarity as the proximity of observations to one another across the variable in the cluster variate. The scaling of the distance measure is such that greater values on this measure denote lesser similarity.

There are several distance measures used during clustering procedures:

- Euclidean distance is the most common distance measure. It is often referred to as a straight line distance.
- Squared Euclidean distance is the sum of the squared differences. It is typically in the Centroid and Wards methods of clustering.
- Manhattan distance uses the sum of absolute differences of the variables. This result may lead to invalid clusters if the variables are highly correlated.
- Clustering algorithms are hierarchical procedures that show similarity is defined between multiple-member clusters. There are numerous approaches to defining similarity between members of each cluster.
- The single linkage or nearest-neighbor method define the similarity between clusters as the shortest distance from any object in one cluster to any object in the others. This method is the most versatile of all agglomerative methods.
- The complete linkage method (or farthest neighbor method) is comparable to the single linkage methods except that the cluster similarity is based on maximum distance between observations in each cluster. This method generates the most compact clustering solutions.
- The average linkage procedure differs from the single linkage or complete linkage procedure in that the similarity of any two clusters is the average similarity of all individuals in one cluster with all individuals in another.
- The centroid linkage method defines the similarity between the two clusters and is the distance between the cluster centroids. Cluster centroids are the mean values of the observations on the variable in the cluster variate. In this method, every time individuals are grouped, a

new centroid is computed. Cluster centroids change every time a new individual or group of individuals is added to the existing clusters.

- The Wards method uses the similarity between the two clusters and is not a single measure of similarity, but rather the sum of squares within the clusters summed over all variables. In this method, the selection of which two clusters to combine is based on which combination of clusters minimizes the within-cluster sum of squares across the complete set of disjoint of separate clusters.

## RESULTS OF THE CLUSTERING ANALYSIS OF ASSET ALLOCATION FUNDS

After an extensive set of clustering analyses using various clustering methodologies (average linkage, centroid linkage, Ward method), distance measures (Manhattan, squared Euclidean, and Euclidean), and cluster sizes, the following three cluster groups were developed, as seen in Table 7.3.

This analysis succeeded in identifying three distinct groups of funds among the 16 different five-star Morningstar-rated moderate asset allocation

**TABLE 7.3 Clusters Underlying Morningstar Classification**

| No. | Fund ID | |
|---|---|---|
| **Panel A: Large Capitalization Funds (Cluster 1)** | | **Consistently Large Fund:** |
| 1 | RLBEX | 7 of 7 |
| 3 | CBAIZ | 7 of 7 |
| 6 | JABAX | 6 of 7 |
| 10 | FOBAX | 7 of 7 |
| 12 | VWELX | 6 of 7 |
| 13 | PRWCX | 7 of 7 |
| 14 | MAPOX | 6 of 7 |
| **Panel B: Small Capitalization Funds (Cluster 2)** | | **Consistently Small Fund:** |
| 5 | FPACX | 7 of 7 |
| 7 | MTOIX | 6 of 7 |
| 8 | PVSAX | 7 of 7 |
| 9 | RNCOX | 6 of 7 |
| 11 | VILLX | 7 of 7 |
| 15 | WBALX | 7of 7 |
| 16 | ACEIX | 6 of 7 |
| **Panel C: Undefined (Cluster 3)** | | **Undefined:** |
| 2 | BUFBX | 3 large; 4 small |
| 4 | HEIBX | 3 large; 5 small |

funds. While these funds are all five-star Morningstar-rated moderate asset allocation funds, they are inherently different, based on a statistical evaluation of the 53 financial variables of the 16 funds. Thus, this analysis presents three sets of investment alternatives. In essence, a screening procedure within the same set of like-rated and classified funds has been established. These results suggest that classifications such as, most prominently, Morningstar's that clump funds within a conjoint classification as five-star moderate asset allocation hide heterogeneous elements or, here, clusters. Understanding the result found here that seeming uniformity hides a hidden heterogeneity is important in developing a better understanding of one's portfolio composition. Even so, other schemes exist that argue for a minute breaking apart of funds, with each fund being evaluated independently. Various organizations, for example, offer screening tools to use in evaluating mutual fund choices. Bloomberg.com (http://www.bloomberg.com/apps/data?pid=fundscreener), for example, provides tools allowing the user to screen mutual funds based on country, asset class focus, fund type, fund size, fees and expenses, and return over different time periods. In a related vein, U.S. News (http://money.usnews.com/funds/mutual-funds) offers mutual fund rankings that utilize expert opinion. It states:

> By compiling expert fund opinion, the *U.S. News* Mutual Fund Score offers a broad look at what some of the brightest minds conducting investing analysis have to say about some of your most important investments. Using ratings from Morningstar, Lipper, Zacks, TheStreet.com, and Standard & Poor's, we've built a one-of-a-kind, one-stop site for following your favorite funds.

The U.S. News site then notes that while raters such as Morningstar analyze historic performance, the U.S. News site utilizes expert opinion to project future performance (see U.S. News Money, 2010, for more detailed discussion). This tool, though, is based on equally weighting the overall ratings provided by its sources. The empirical basis for such a weighting scheme is not provided. Accordingly, it seems to have been established a priori. To the extent that the underlying sources have problematic methodologies and the classification of different variables (e.g., risk-adjusted historical returns) differs between the sources, even a weighted aggregation of the weighted items may yield further error.

Tools such as this honor differences between funds but do not provide easy means of evaluating the commonality among funds. The classification tool used in this study enables isolation of both the commonality and sources of difference between funds. In doing so, it provides superior insight into the commonalities and differences within a conjoint categorization of funds that may have been regarded as homogeneous. A limitation of this study is that it examines only one conjoint classification of Morningstar funds. The tools utilized here, though, can be employed to examine whether similar heterogeneity exists within other conjoint Morningstar classifications. It may also be employed in examining underlying heterogeneity in the conjoint classifications of other assets.

## REFERENCES

Barron, J. M., & Ni, J. (2013). Morníngstar ratings and mutual fund manager turnover. *Journal of Applied Finance, 1,* 95–110.

Carlson, G. (2013, August 6). Dodge & Cox, Vanguard, T. Rowe Price, and more: Are these funds' recent returns sustainable? *Morningstaar Newsletter,* p. 1.

Morningstar (2010). *The Morningstar rating for funds fact sheet.* Retrieved from http://corporate.morningstar.com/documents/MethodologyDocuments/FactSheets/MorningstarRatingForFunds_FactSheet.pdf

Otranto, E. (2008). Clustering heteroskedastic time series by model-based procedures. *Computational Statistics, 52,* 4685–4698.

Paterlin, F., Paterlini, S., & Minerva, T. (2004). Clustering financial time series: An application to mutual funds style analysis. *Computational Statistics & Data Analysis, 47,* 353–372.

Perez-Gladish, B., Rodriguez, P. M., M'Zali, B., & Lang, P. (2013). Mutual funds efficiency measurement under financial and social responsibility criteria. *Journal of Multi-Criteria Decision Analysis, 20,* 109–125.

Sharpe, W. F. (1992). Asset allocation: Management style and performance measures. *Journal of Portfolio Management, 18,* 7–19.

Sharpe, W. F. (1998, July/August). Morningstar's risk-adjusted ratings. *Financial Analysts' Journal, 54*(4), 21–33.

Thune, K. (2013). *Mutual fund categories: How funds are categorized.* Retrieved from http://mutualfunds.about.com/od/typesoffunds/a/Mutual-Fund-Categories.htm

U.S. News Money. (2010, February 23). *About the U.S. News mutual fund score.* Retrieved from http://money.usnews.com/money/personal-finance/investing/articles/2010/02/23/about-the-us-news-mutual-fund-score

# SECTION III

## TECHNIQUES

CHAPTER 8

# DATA MINING TECHNIQUES APPLIED TO THE STUDY OF CANINES WITH OSTEOARTHRITIS

## Developing a Predictive Model

**Virginia M. Miori**
***St. Joseph's University***

## ABSTRACT

This is the second publication in a series aimed at providing models effective in predicting the degree of pain and discomfort in canines. The first paper presented forecasting models aimed at predicting activity levels in normal (control group) dogs. In this chapter, the focus shifts to the evaluations of data drawn from both normal dogs and dogs suffering from osteoarthritis: demographic data, normal activity data, pretreatment activity data, and post-treatment activity data. Microsoft Access and SAS JMP tools are used to generate ANOVA, LSR regression, decision tree, and nominal logistic regression models to predict changes in activity levels associated with the progression of arthritis. The predictive models provide a diagnostic basis for determining

*Contemporary Perspectives in Data Mining, Volume 2*, pages 139–175

the degree of disease in a dog (based on demographics and activity levels) and provide information that can assist in effective diagnosis and medication of suffering dogs.

## INTRODUCTION

In a previous paper, activity levels for normal dogs were forecasted based on demographic characteristics of the dogs (age, weight, and sex). Though these forecasts were statistically accurate, they were created using ARIMA (autoregressive integrated moving average) models, limiting their applicability. They did, however, offer a reasonable expectation that additional data could be used in finding improved and more practical prediction methods.

The current research includes data collected through several studies at the University of Pennsylvania. The studies included normal dogs and dogs suffering from osteoarthritis. Data represented untreated and medicated phases of the study for the osteoarthritic dogs. The demographics have been extended from the previous factors to include pain survey responses, function survey responses, and typical daily levels of interaction between owners and dogs. Activity data were collected every sixty seconds, over all phases of study, using Actical monitors. In all, 26 normal dogs and 71 dogs with osteoarthritis were studied. The average length of a study phase was fourteen days, and each dog participated in two to three phases resulting in approximately six million records of activity data.

The ultimate goal of this research is to provide a diagnostic tool to predict the level of pain/discomfort being experienced by a dog. This leads to more precise dosing of medication, ensuring that dogs are not improperly medicated. In cases of over-medication, the side effects of medications can be harmful to the dogs overall health. Under-medication is undesirable because it would leave a dog in continued pain or discomfort.

## LITERATURE

The literature is presented in discrete sections, each addressing one aspect of the chapter. The first section addresses the validation of the use of activity monitors, originally designed for human use, in dogs. The second section addresses studies in which the data from activity monitors have been used in evaluations of pain and discomfort. These studies have also validated appropriate data collection intervals. The use of activity monitors on dogs is a recent development resulting in limited research in the specifically related areas. The final section examines various data mining techniques that may be useful in predicting discomfort due to arthritis.

There is a dearth of dog-related literature in this area; therefore, studies of predicting human arthritis are referenced along with more general aspects of predicting dog health.

## Validate Activity Monitors

The omnidirectional accelerometer-based devices used to record activity levels in dogs were Actical Activity Monitors (AAM) and were initially produced for use in humans by the Respironics Mini Mitter division in Bend, Oregon. The monitors were watch-sized, placed in waterproof housings, and attached to the collars of the dogs. The AAMs continuously recorded intensity, frequency, and duration of movement and were set to record data at 60-second epochs (intervals), 24 hours every day.

Validation of the use of activity monitors in dogs was first studied in conjunction with videographic measurements of movement and mobility in healthy dogs (Hansen, Lascelles, Keene, Adams, & Thomson, 2007). Monitors were placed at locations on the dogs to test their output in order to determine the most effective location. Movement and mobility of dogs were recorded with a computerized videography system for seven-hour sessions while the dogs were also wearing activity monitors. Accelerometer values were combined into 15-minute intervals and compared with videographic measures (distance traveled, time spent walking, and time spent changing position). When comparing accelerometers with the greatest disparity, 96% of all values compared were within two standard deviations of the mean. All monitor placement locations provided acceptable correlation with the videographic data, but the ventral collar (front of the dog's neck) was determined to be most convenient.

Once activity monitors were determined to provide accurate results, the optimal or most appropriate sampling interval was evaluated (Dow, Michel, Love, & Cimino Brown, 2009). Dogs in this study wore monitors for two weeks. Variation between dogs and within dogs were evaluated on a day-to-day basis and then evaluated between week 1 and week 2, comparing for weekdays, weekends, and full weeks. Significant variation in activity counts between dogs was detected as well as variation within dogs over the course of a full week. The study determined that monitors should be word for seven-day intervals in order to follow activity over time.

Body conformation and signalment (medical history dealing with a dog's age, sex, and breed; Signalment, n.d.) were examined in the use of activity monitors (Cimino Brown, Michel, Love, & Dow, 2010). A sample of 104 companion dogs was led through a series of standard activities to determine whether signalment and body conformation had an impact on activity counts. It was determined with 95% confidence that activity counts decreased with increased body weight. Activity counts also decreased with increases in age.

The GT3-X accelerometer has been studied for validity as well as practical utility and reliability in measuring activity in pets (Yam et al., 2011). The validation study examined 30 dogs for one day, filming them while they wore the monitors. Data and film were synchronized to reflect variations in activity levels of a habitual nature. Reliability and practical utility were evaluated on 20 dogs who wore the monitors for one week. The data were supplemented by owner questionnaires. In support of validity, activity levels differed significantly and compared directly to activity intensity. The devices were also found to be well tolerated (practical utility) and to have minimal data loss (reliability), supporting the conclusions that the devices are a valid, practical, and reliable way to collect habitual activity data in dogs.

Differentiation among activities of differing intensities and delineation of the times spent in these activities was completed in a two-phase study (Michel & Cimino Brown, 2011). During the first phase, dogs were led through a series of activities at differing levels of intensity. Characteristic curves were developed and used to determine the most prevalent activity counts when predicting the intensity of these activity levels. Dogs wore the monitors at home for two weeks during the second phase of the study. The outcome of the first phase was used to classify intensity of activity during the 14-day period. Distinguishing sedentary activity from walking activity was highly significant, as was distinguishing trotting activity from walking activity.

The cumulative outcomes of the validity studies have strongly supported the use of activity monitors in assessing activity levels in dogs at varied levels of intensity of activity.

## Studies Utilizing Activity Monitors

Once the use of activity monitors was determined to be effective, examining treatment responses followed. Researchers had found a way to evaluate levels of pain and discomfort in dogs that could be used to support other anecdotal and clinical measurements.

Discomfort in dogs takes many forms; pruritus (itching) in dogs is one such form of discomfort. It impacts quality of life and is difficult to assess objectively. Atopic dermatitis in dogs was studied using Actiwatch® collar-mounted activity monitors (Nuttall & McEwan, 2006). Activity levels of five normal dogs were compared to six dermatitis dogs, with controls for defined periods of exercise, playing, and so on. Data were collected over seven days, with 15-second epochs. Overall, interquartile ranges for daytime activity were similar during the day and evening, while lower during the night. Using the Mann-Whitney test, the authors found that mean activity during the epochs was significantly higher in atopic dogs, compared to healthy dogs during all three periods.

Activity monitors have been used to compare differences in activity level of dogs undergoing three laparoscopic surgical techniques (LAG, hand suture TLG, and Endostitch™ TLG) in preventative treating of gastric dilation and volvulus (Mayhew & Brown, 2009). This condition is commonly referred to as bloat and occurs most frequently in large and giant breed dogs. The surgical procedure attaches the dog's stomach (gastric antral wall) to their right body wall to prevent rotation. Monitors were placed on dogs for seven days before surgery and seven days after surgery. The activity counts gathered were used to compare recovery rates of dogs between the two laparoscopic procedures. The result of the analysis showed LAG to result in more greatly reduced activity levels than either TLG procedure. No differences in activity levels were demonstrated between TLG techniques.

A similar study examined activity levels in small (<10 kg) female dogs undergoing sterilization (Culp, Mayhew, & Brown, 2009). A laparoscopic surgical procedure (LapOVE) was compared to the traditional open procedure (OOVE). Activity data were collected 24 hours prior to surgery and 48 hours after surgery, with the removal and return of the dog to her housing run marking the bounds for collection of data. As expected, activity levels in dogs in the OOVE group were substantially lower (62% decrease compared to baseline) than those in the LapOVE group (25% decrease compared to baseline).

Evaluation of osteoarthritis treatment in dogs has traditionally been completed by the use of gait analysis and descriptions of activity provided by owners. Activity monitors have offered an alternative to descriptive data and gait analysis in evaluating the efficacy of medications. A study of 70 dogs with osteoarthritis was completed to evaluate the efficacy of carprofen (Cimino Brown, Boston, & Farrar, 2010). Dogs were monitored for 21 days; no treatment was given for the first 7 days. During days 8–21, dogs were either given a placebo (control group) or carprofen. Changes in activity level were evaluated between the first 7 days and the remaining 14 days for each group. Linear regression was used to determine associations between treatments and percentage change in activity counts while controlling for conformation and signalment variables. With 95% confidence, an increase of 20% in median activity count was found, suggesting that carprofen was effective in treating osteoarthritis.

Canine hyperactivity (hyperkinesis syndrome, a.k.a HS) has been studied in Beagles. Accelerometers have been used to assist in determining the effectiveness of dextroamphetamines in treatment of HS (Stiles, Palestrini, Beauchamp, & Frank, 2011). In dogs with HS, oral dextroamphetamines have demonstrated a paradoxical effect and therefore do not elevate activity levels. This study demonstrated the same findings with lower doses. Beagles wore collar-mounted AAMs and recorded 180 minutes of activity recorded at 15-second epochs. Monitor data were compared to video collected during the same time and found to be consistent with the videos. Results also showed no significant effects of treatment on the dogs' activity level.

Accelerometry was used in an assessment of the ability to reduce doses of nonsteroidal anti-inflammatory drugs (NSAID) in dogs, yet still control pain resulting from osteoarthritis (Wernham et al., 2011). A number of dogs dropped out of the study because their owners determined that the pain control was insufficient. Among the remaining 59 dogs, activity monitoring was maintained during the duration of the study. Posthoc Bonferoni analysis was used to determine that there was no evidence of effects for the percentage of time above upper thresholds for activity.

The final study in this area has the specific purpose of measuring daily activity, using activity monitors, in health adult Labrador retrievers as a predictor of maintenance energy requirements (MER) (Wrigglesworth, Mort, Upton, & Miller, 2011). Dogs wore the monitors for a two-week period. Activity counts in conjunction with data on daily activity levels and body weight were characterized as independent variables in a multiple linear regression to predict daily MER. A second regression excluding activity counts was also completed. Dietary energy intake at a stated body weight was used as a proxy variable for MER. Inclusion of the activity levels significantly improved the predictive capabilities of the multiple linear regression model.

Activity monitors proved to be useful and significant in studying various conditions and determining energy needs for dogs. As this area of study expands, it is important to develop a method of prediction for baseline activity levels in dogs. This baseline may then be used to further assess the impact of pain and discomfort on activity levels.

## Predictive Technique for Arthritis

The nonparametric $\chi^2$ technique was used to examine differences in personality associated with human rheumatoid arthritis patients at different stages of the disease (Ward, 1971). Personality surveys were used to supplement diagnostic data in arthritis patients, normal control group patients, and neurotic patients.

Genetic factors associated with rheumatoid arthritis were used to predict the development of radiological erosions in patients with early symmetrical (rheumatoid-like) arthritis (Emery et al., 1992). The Mann-Whitney U test was used along with confidence intervals to produce predictions of incidence. Logistic regression was employed within a simple heuristic to predict the development of radiological erosions in patients with rheumatoid arthritis (Brennan et al., 1996). Explanatory variables were patient characteristics and patient condition, with ultimate identification of three significant variables.

Smoking, obesity, and alcohol consumption were studied as risk factors for increased incidence of rheumatoid arthritis (Voigt, Koepsell, Nelson, Dugowson, & Daling, 1994). The analysis used confidence intervals and odds ratios to quantify the increased risk of arthritis for each of the three factors as well as the impact of menopause on these factors.

A bootstrap method with Monte Carlo resampling was used to predict morbidity of acute diseases (Alonso & Romo, 2005). This paper sought to move away from ARIMA models for more accurate time series predictions. Frequent and prompt resampling was performed to maintain high levels of accuracy in reporting and determination of bootstrap prediction intervals.

Joint damage in psoriatic arthritis patients is typically evaluated through clinical assessment and through X-rays of joints. Clinical assessments are more subjective, but due to safety of the patients and the need to control healthcare costs, X-rays are taken (approximately) every two years. Negative binomial regression was used to more accurately evaluate patient condition during times of clinical evaluation (Bond & Farewell, 2009).

## DATA OVERVIEW

Data were collected on dogs in a series of studies performed at the University of Pennsylvania. This included extensive demographic data and activity count data.

The demographic data on each dog were matched to the appropriate activity count data. The demographic data included age, weight, sex, breed, and treatment phase for each dog. In addition, owners completed a number of surveys. Though significantly more demographic data were available for the dogs in different studies, this research used only survey data that precisely matched between the normal dogs and the osteoarthritic dogs. This included a brief pain inventory survey reflecting observations of pain over the previous ten days, a level of interaction survey, also reflecting observations of the previous ten days, and a description of function reflecting observations over the previous seven days. The surveys are included in Appendix A of this chapter.

Activity data have been collected using Actical® activity monitors. These monitors were designed for human use but have been proven effective for measuring activity in dogs when attached to the dog's collar. The values collected with these monitors are indications of electrochemical activity. They provide comparative values but have no direct intuitive interpretation. Most dogs in the study wore the monitors for six weeks, though a few wore them for only four weeks. During this time, activity counts were collected every 60 seconds, resulting in approximately six million records of activity data.

Rather than attempt to complete detailed analysis of these six million records, a logical aggregation of the data has been completed. Data cleaning was necessary to match activity data to demographic data. Study phase names and dog references were inconsistent within and across studies. Manual examination of all records was required to ensure consistency in data aggregation and data matching.

An average activity count was calculated for every minute of the day, for each dog, for each phase of the study. Once the aggregation was completed, the demographic data were matched to the average activity records in Microsoft Access. These data were exported and served as the base for all analyses presented in this chapter.

Overall data patterns were examined to confirm that expectations of activity levels were met. A graph showing cumulative averages of activity levels for all normal dogs (Normal), untreated osteoarthritic dogs (No), and treated osteoarthritic dogs (Yes) is presented in Figure 8.1. Untreated dogs who experience stiffness are expected to demonstrate erratic and higher levels of activity. This is due to overall discomfort resulting in an inability to remain still. Once treated, the osteoarthritic dogs are expected to exhibit more consistent and lower levels of activity. Normal dogs are expected to fall in between these two extremes. This expected pattern of activity is clearly displayed in Figure 8.1.

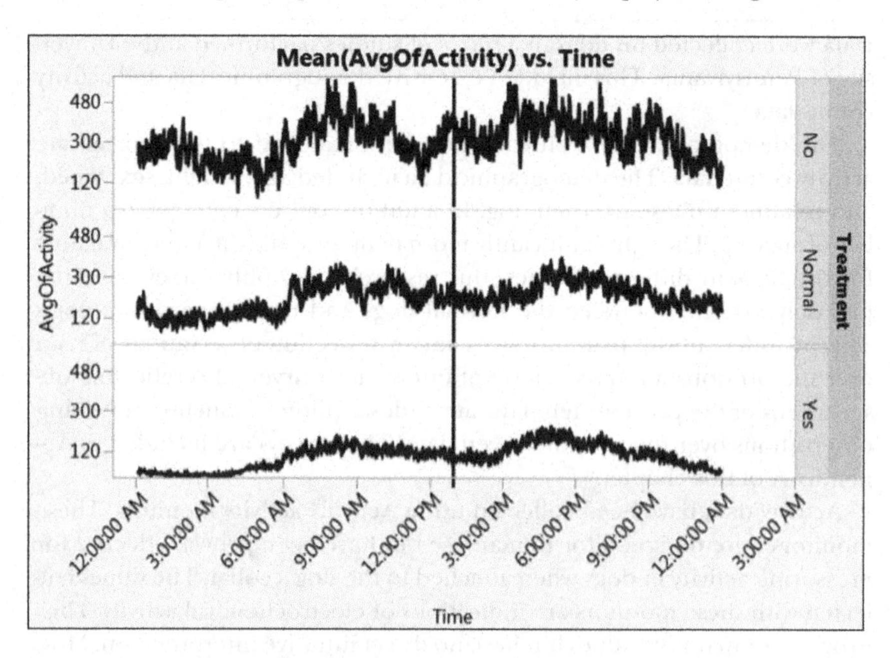


**Figure 8.1** Comparative activity levels between dog study phases.

## ANALYSIS

The goal of the analyses was to produce a consistent predictive approach. This approach not only had the purpose of predicting the average activity level, but additionally to establish a significant relationship and identify patterns between average activity level and treatment.

All analyses were completed using JMP software by SAS. Analyses of variance, nominal logistic regression, least squares regression, and decision trees were the predominate tools leading to the most promising results. Dozens of models were run; the results below show the most promising of these analyses.

### ANOVA

ANOVA models provided strong results that confirmed the initial relationships between demographic factors and the average activity level. Due to the fact that surveys were the origin of many of the data points, ANOVAs with more than one factor have consistently resulted in extensive singularities. These ANOVAs are not presented in this chapter due to the violation of ANOVA assumptions. These variables did, however, prove significant in a series of single factor ANOVAs examining differences in average activity level of the dogs. Table 8.1 contains a summary of these results, in order of decreasing significance.

**TABLE 8.1 One Factor ANOVA Summary**

| Variable | Adjusted R Square | p-value |
|---|---|---|
| Breed | 0.261 | <0.0001 |
| Pain on Rising | 0.070 | <0.0001 |
| Pain while Running | 0.052 | <0.0001 |
| Pain Interferes with Enjoyment of Life | 0.041 | <0.0001 |
| Age | 0.038 | <0.0001 |
| Daytime Level of Interaction | 0.036 | <0.0001 |
| Pain at its Worst | 0.023 | <0.0001 |
| General Level of Pain | 0.019 | <0.0001 |
| Level of Active Interaction | 0.018 | <0.0001 |
| Pain while Walking | 0.016 | <0.0001 |
| Pain while Climbing | 0.016 | <0.0001 |
| Current Pain | 0.014 | <0.0001 |
| Average Pain | 0.013 | <0.0001 |

Differences between breeds were found to be most significant, accounting for over 26% of the variation in activity level. Pain on rising, pain while running, pain interfering with enjoyment of life, age, and daytime levels of interactions were collectively found to account for almost 24% of the variation in activity level. ANOVA models were not reported for variables resulting in the adjusted R square falling below 1.2%. The JMP output for the breed ANOVA is contained in Appendix B. All other outputs are available upon request.

Three-factor ANOVA provided support for significance of interactions between age, weight, and treatment. As expected, the two-way interactions between all of these factors were significant. The JMP output for this ANOVA is found in Appendix B.

Multifactor ANOVA generated strong results with characteristic data for the dogs. The characteristics most often (anecdotally) associated with changes in physical condition are treatment, sex, age, weight, and breed. The ANOVA summary is contained in Table 8.2, while the effects test is contained in Table 8.3. All of the factors tested were found to be highly significant in predicting average activity levels. Parameter estimates and summary of fit are found in Appendix B.

Further examination of results, using the Tukey HSD (honest significant difference) test, identified cases in which pairwise differences exist between categories of each variable. Tables 8.4 and 8.5 provide definitions of the age and weight grouping variables. (Note that this dataset had no dogs

**TABLE 8.2 Multi Factor ANOVA**

| | Analysis of Variance | | | |
|---|---|---|---|---|
| **Source** | **DF** | **Sum of Squares** | **Mean Square** | **F Ratio** |
| Model | 34 | 1.1277e+11 | 3.3168e+9 | 4137.979 |
| Error | 370045 | 2.9661e+11 | 801539.77 | **Prob > F** |
| C. Total | 370079 | 4.0938e+11 | | <.0001* |

**TABLE 8.3 Multi Factor Effects Test**

| | Effect Tests | | | | |
|---|---|---|---|---|---|
| **Source** | **Nparm** | **DF** | **Sum of Squares** | **F Ratio** | **Prob > F** |
| Treatment | 2 | 2 | 4067990661 | 2537.610 | <.0001* |
| Age Group | 5 | 5 | 798490034 | 199.2390 | <.0001* |
| SexD | 1 | 1 | 91580150.1 | 114.2553 | <.0001* |
| Weight Group | 4 | 4 | 828503944 | 258.4101 | <.0001* |
| Breed | 22 | 22 | 9.0916e+10 | 5155.762 | <.0001* |

**TABLE 8.4 Age Groupings**

| Age Group Number | Ages Represented |
|---|---|
| 1 | 0 up to 4 years |
| 2 | 4 up to 6 years |
| 3 | 6 up to 8 years |
| 4 | 8 up to 10 years |
| 5 | 10 up to 12 years |
| 6 | 12 years and older |

**TABLE 8.5 Weight Groupings**

| Weight Group Number | Weights Represented |
|---|---|
| 1 | 0 up to 10 kg |
| 2 | 10 up to 20 kg |
| 3 | 20 up to 30 kg |
| 4 | 30 up to 40 kg |
| 5 | 40 up to 50 kg |
| 6 | 50 kg and heavier |

weighing less than 10 kg.) Differences were identified between treatment categories in Table 8.6, between age categories in Table 8.7, and between weight categories in Table 8.8. The significance of these differences supports inclusion of these variables in logistic and least squares regression.

No significance was attached to differences in activity between age groups 5 (10 up to 12 years) and 6 (12 years and over). Additionally, no differences were significant between weight groups 2 (10 up to 20 kg) and 3 (20 up to 30 kg), weight groups 2 (10 up to 20 kg) and 5 (40 up to 50 kg), and weight groups 3 (20 up to 30 kg) and 5 (40 up to 50 kg).

**TABLE 8.6 Pairwise Differences between Treatments**

| Treatment | -Treatment | Difference | Std Error | DF | t Ratio | Prob>\|t\| | Lower 95% | Upper 95% |
|---|---|---|---|---|---|---|---|---|
| No | Normal | 126.4952 | 5.634129 | 370045 | 22.45 | <.0001* | 113.2904 | 139.6999 |
| No | Yes | 237.0675 | 3.328556 | 370045 | 71.22 | <.0001* | 229.2664 | 244.8687 |
| Normal | Yes | 110.5724 | 5.495441 | 370045 | 20.12 | <.0001* | 97.6926 | 123.4521 |

**TABLE 8.7 Pairwise Differences between Age Groups**

| Age Group | -Age Group | Difference | Std Error | DF | t Ratio | Prob>\|t\| | Lower 95% | Upper 95% |
|---|---|---|---|---|---|---|---|---|
| 1 | 2 | –85.950 | 6.75853 | 370045 | –12.72 | <.0001* | –105.210 | –66.690 |
| 1 | 3 | 34.670 | 7.59742 | 370045 | 4.56 | <.0001* | 13.019 | 56.320 |
| 1 | 4 | 58.224 | 6.41719 | 370045 | 9.07 | <.0001* | 39.937 | 76.511 |
| 1 | 5 | 117.090 | 6.94384 | 370045 | 16.86 | <.0001* | 97.302 | 136.878 |
| 1 | 6 | 116.838 | 9.87754 | 370045 | 11.83 | <.0001* | 88.690 | 144.987 |
| 2 | 3 | 120.620 | 6.82182 | 370045 | 17.68 | <.0001* | 101.180 | 140.060 |
| 2 | 4 | 144.174 | 5.67808 | 370045 | 25.39 | <.0001* | 127.993 | 160.355 |
| 2 | 5 | 203.040 | 6.72693 | 370045 | 30.18 | <.0001* | 183.871 | 222.210 |
| 2 | 6 | 202.789 | 9.90572 | 370045 | 20.47 | <.0001* | 174.560 | 231.017 |
| 3 | 4 | 23.555 | 6.05916 | 370045 | 3.89 | 0.0014* | 6.288 | 40.821 |
| 3 | 5 | 82.421 | 6.65712 | 370045 | 12.38 | <.0001* | 63.450 | 101.392 |
| 3 | 6 | 82.169 | 10.08591 | 370045 | 8.15 | <.0001* | 53.427 | 110.911 |
| 4 | 5 | 58.866 | 4.87136 | 370045 | 12.08 | <.0001* | 44.984 | 72.748 |
| 4 | 6 | 58.614 | 8.94634 | 370045 | 6.55 | <.0001* | 33.120 | 84.109 |
| 5 | 6 | –0.252 | 8.92901 | 370045 | –0.03 | 1.0000 | –25.697 | 25.193 |

**TABLE 8.8 Pairwise Difference between Weight Groups**

| Weight Group | -Weight Group | Difference | Std Error | DF | t Ratio | Prob>\|t\| | Lower 95% | Upper 95% |
|---|---|---|---|---|---|---|---|---|
| 2 | 3 | 2.003 | 7.40298 | 370045 | 0.27 | 0.9988 | –18.190 | 22.197 |
| 2 | 4 | –126.009 | 8.13608 | 370045 | –15.49 | <.0001* | –148.203 | –103.816 |
| 2 | 5 | –0.708 | 10.09630 | 370045 | –0.07 | 1.0000 | –28.249 | 26.832 |
| 2 | 6 | 41.204 | 10.74667 | 370045 | 3.83 | 0.0012* | 11.890 | 70.519 |
| 3 | 4 | –128.013 | 4.68090 | 370045 | –27.35 | <.0001* | –140.781 | –115.244 |
| 3 | 5 | –2.712 | 7.02672 | 370045 | –0.39 | 0.9953 | –21.879 | 16.456 |
| 3 | 6 | 39.201 | 8.15841 | 370045 | 4.80 | <.0001* | 16.947 | 61.455 |
| 4 | 5 | 125.301 | 6.75264 | 370045 | 18.56 | <.0001* | 106.881 | 143.721 |
| 4 | 6 | 167.214 | 7.94503 | 370045 | 21.05 | <.0001* | 145.541 | 188.886 |
| 5 | 6 | 41.913 | 8.46123 | 370045 | 4.95 | <.0001* | 18.832 | 64.993 |

## Nominal Logistic Regression

The logistic regression capability in JMP extends to count data analysis, allowing discrete response variables, not just binary response variables. Treatment was designated as the response variable, and after a series of nominal logistic runs, the significant (nonsingular) variables were identified. These variables:

- age
- weight

- general pain index
- all interactions between these variables

resulted in a generalized R square of 0.5965.

The logistic regression is useful for predicting the level of treatment in a particular dog. This is a useful tool in affirming the need for treatment in any given dog; however, the ability to predict the level of activity in a dog and the identification of the contributions of survey factors are critical to diagnosis of dogs. Least squares regression with average activity level as a response variable more adequately served this purpose. The complete output for the nominal logistic regression is available in Appendix C.

### Least Squares Regression

A series of least square regression (LSR) models ultimately allowed for identification of the following significant variables:

- time of day
- treatment
- age
- sex
- weight
- breed
- pain
  - at its worst
  - at its least
  - general
  - interferes with enjoyment of life
  - on rising
  - on walking
  - on running
  - on climbing
  - interferes with overall activity
- lifestyle
- days with owner
- hours awake with owner
- hours of active interaction
- nights slept in same room as owner
- functional problems
  - rising
  - climbing up
  - climbing down

- jumping up
- jumping down
- posturing to urinate or defecate

The adjusted R square for the resulting LSR was 0.6904, indicating a strong model, and the F test was very significant. These results are contained in Tables 8.9 and 8.10. All variables were strongly significant with the exception of one dummy variable within breed, one dummy variable within pain on walking, and one dummy variable within lifestyle. These insignificant dummy variables showed that Akitas were not distinct in breed, high levels of interference due to pain when walking do not differ, and no difference exists between dogs who spend all of their time outdoors and dogs that spend time both inside and outdoors. The effects tests are provided in Table 8.11. A complete summary of the least squares regression is available in Appendix D.

The LSR has provided the strongest diagnostic tool. If used in conjunction with the nominal logistic regression, it will provide both a starting point for effective prescribing of medication and an expectation of the improvements that can be garnered due to medication.

## Decision Tree

The final examination of data was completed using decision trees. Though the decision tree was subject to issues of singularity, these issues were determined to be coincidental rather than programmatic in nature. Selected aspects of the decision tree results are presented in Figures 8.2, 8.3, and 8.4. Figure 8.2 shows a misclassification rate below 5%. The confusion matrix in

**TABLE 8.9 Summary of Fit**

| | |
|---|---|
| RSquare | 0.69049 |
| RSquare Adj | 0.690358 |
| Root Mean Square Error | 593.1639 |
| Mean of Response | 213.3081 |
| Observations (or Sum Wgts) | 360000 |

**TABLE 8.10 Analysis of Variance**

| Source | DF | Sum of Squares | Mean Square | F Ratio |
|---|---|---|---|---|
| Model | 153 | 2.8245e+11 | 1.8461e+9 | 5246.958 |
| Error | 359846 | 1.2661e+11 | 351843.43 | Prob > F |
| C. Total | 359999 | 4.0906e+11 | | <.0001* |

**TABLE 8.11 Effects Tests**

| Source | Nparm | DF | Sum of Squares | F Ratio | Prob > F |
|---|---|---|---|---|---|
| Time | 1 | 1 | 266448775 | 757.2936 | <.0001* |
| Treatment | 2 | 2 | 7762158668 | 11030.70 | <.0001* |
| AgeD | 1 | 1 | 269008394 | 764.5685 | <.0001* |
| SexD | 1 | 1 | 991106622 | 2816.897 | <.0001* |
| Weight | 1 | 1 | 8749916.85 | 24.8688 | <.0001* |
| Breed | 22 | 22 | 1.0508e+11 | 13575.39 | <.0001* |
| BPI_PainWst | 10 | 10 | 4978472754 | 1414.968 | <.0001* |
| BPI_PainLst | 8 | 8 | 4178060714 | 1484.347 | <.0001* |
| BPI_General | 10 | 10 | 5418816333 | 1540.122 | <.0001* |
| BPI_Life | 9 | 9 | 6306647261 | 1991.621 | <.0001* |
| BPI_Rise | 10 | 10 | 8746871940 | 2486.013 | <.0001* |
| BPI_Walking | 9 | 9 | 6010408510 | 1898.069 | <.0001* |
| BPI_Run | 10 | 10 | 6848088556 | 1946.345 | <.0001* |
| BPI_Climb | 10 | 10 | 3362856639 | 955.7821 | <.0001* |
| BPI_Overall | 4 | 4 | 762535418 | 541.8145 | <.0001* |
| LOI_Lifestyle | 4 | 4 | 9306528648 | 6612.692 | <.0001* |
| LOI_Days | 4 | 4 | 2.4074e+10 | 17105.68 | <.0001* |
| LOI_Hours | 3 | 3 | 617471660 | 584.9872 | <.0001* |
| LOI_Interact | 4 | 4 | 937919338 | 666.4323 | <.0001* |
| LOI_Nights | 7 | 7 | 3474796215 | 1410.853 | <.0001* |
| _7Func Rise | 4 | 4 | 2126284548 | 1510.817 | <.0001* |
| 11FuncClDo | 4 | 4 | 328839303 | 233.6546 | <.0001* |
| 10FuncClUp | 4 | 4 | 3723516406 | 2645.720 | <.0001* |
| 8FuncJuUp | 4 | 4 | 1997551968 | 1419.347 | <.0001* |
| 9FuncJuDo | 4 | 4 | 1720578839 | 1222.546 | <.0001* |
| 12FuncPost | 3 | 3 | 2482143544 | 2351.561 | <.0001* |

**Fit Details**

| Measure | Training | Definition |
|---|---|---|
| Entropy RSquare | 0.9055 | 1-Loglike(model)/Loglike(0) |
| Generalized RSquare | 0.9691 | (1-(L(0)/L(model))^(2/n))/(1-L(0)^(2/n)) |
| Mean -Log p | 0.0984 | ∑ -Log(ρ[j])/n |
| RMSE | 0.1802 | $\sqrt{\sum(y[j]-\rho[j])^2/n}$ |
| Mean Abs Dev | 0.0650 | ∑ \|y[j]-ρ[j]\|/n |
| Misclassification Rate | 0.0465 | ∑ (ρ[j]≠ρMax)/n |
| N | 371520 | n |

**Figure 8.2** Decision tree fit details.

**Confusion Matrix**

| Actual \ Predicted: Training | No | Normal | Yes |
|---|---|---|---|
| No | 142560 | 0 | 12960 |
| Normal | 0 | 67680 | 0 |
| Yes | 4320 | 0 | 144000 |

**Figure 8.3** Decision tree confusion matrix.

**Column Contributions**

| Term | Number of Splits | G^2 | Portion |
|---|---|---|---|
| BPI_PainWst | 4 | 274794.554 | 0.3922 |
| Breed | 5 | 92674.3049 | 0.1323 |
| BPI_Rise | 4 | 67190.2954 | 0.0959 |
| BPI_Climb | 3 | 60779.6459 | 0.0867 |
| BPI_Run | 3 | 47202.5132 | 0.0674 |
| BPI_PainNow | 1 | 32921.5354 | 0.0470 |
| BPI_Life | 3 | 24913.8456 | 0.0356 |
| BPI_General | 2 | 18263.2348 | 0.0261 |
| LOI_Interact | 1 | 16837.8975 | 0.0240 |
| Weight | 1 | 16258.8393 | 0.0232 |
| BPI_Walking | 1 | 15331.3571 | 0.0219 |
| LOI_Nights | 1 | 14400.4898 | 0.0206 |
| 11FuncClDo | 1 | 10354.8024 | 0.0148 |
| 8FuncJuUp | 1 | 8764.31614 | 0.0125 |

**Figure 8.4** Decision tree column contributions.

Figure 8.3 shows the number of misclassified observations. Finally, the column contribution presented in Figure 8.4 validates the variables found to be significant in the LSR. Complete analysis results are presented in Appendix E.

## CONCLUSIONS

The extended analysis performed in this chapter using ANOVA, logistic regression, least squares regression, and decision trees has demonstrated definitive patterns in the demographic and activity level data. The most significant factors in predicting osteoarthritis were confirmed to be age, weight, observation of pain and function, and level of owner interaction with the dogs. Determining the contribution of each of these factors was a critical outcome of the analyses.

The finalized models must now be put into use for diagnostic purposes. Extended study, in which data are collected on the efficacy of these models, is required to further refine their use.

## REFERENCES

Alonso, A. M., & Romo, J. J. (2005). Forecast of the expected non-epidemic morbidity of acute diseases using resampling methods. *Journal of Applied Statistics, 32*(3), 281–295.

Brennan, P., Harrison, B., Barrett, E., Chakraverty, K., Scott, D., Silman, A., & Symmons, D. (1996). Simple algorithm to predict the development of radiological erosions in patients with early rheumatoid arthritis. *The British Medical Journal, 313*(7055), 471–476.

Bond, S. J., & Farewell, V. T. (2009). Likelihood estimation for a longitudinal negative binomial regression model with missing outcomes. *Journal of the Royal Statistical Society. Series C (Applied Statistics), 58*(3), 369–382.

Cinimo Brown, D., Boston, R. C., & Farrar, J. T. (2010). Use of an activity monitor to detect response to treatment in dogs with osteoarthritis. *Journal of the American Veterinary Medical Association, 237*(1), 66–70.

Cimino Brown, D., Michel, K. E., Love, M., & Dow, C. (2010). Evaluation of the effect of signalment and body conformation on activity monitoring in companion dogs. *American Journal of Veterinary Research, 71*(3), 322–325.

Culp, W. T. N., Mayhew, P. D., & Cimino Brown, D. (2009). The effect of laparoscopic versus open ovariectomy on postsurgical activity in small dogs. *Veterinary Surgery, 38*, 811–817.

Dow, C., Michel, K. E., Love, M., & Cimino Brown, D. (2009). Evaluation of optimal sampling interval for activity monitoring in companion dogs. *American Journal of Veterinary Research, 70*(4), 444–448.

Emery, P., Salmon, M., Bradley, H., Wordsworth, P., Tunn, E., Bacon, P. A., & Waring, R. (1992). Genetically determined factors as predictors of radiological change in patients with early symmetrical arthritis. *The British Medical Journal, 305*(6866), 1387–1389.

Hansen, B. D., Lascelles, B. D. X., Keene, B. W., Adams, A. K., & Thomson, A. E. (2007). Evaluation of an accelerometer for at-home monitoring of spontaneous activity in dogs. *American Journal of Veterinary Research, 68*(5), 468–475.

Mayhew, P. D., & Cimino Brown, D. (2009). Prospective evaluation of two intracorporeally sutured prophylactic laparoscopic gastropexy techniques compared with laparoscopic-assisted gastropexy in dogs. *Veterinary Surgery, 38*, 738–746.

Michel, K. E., & Cimino Brown, D. (2011). Determination and application of cut points for accelerometer-based activity counts of activities with differing intensity in pet dogs. *American Journal of Veterinary Research, 72*(7), 866–870.

Nutall, T., & McEwan, N. (2006). Objective measurement of pruritus in dogs: A preliminary study using activity monitors. *European Society of Veterinary Dermatology, 17,* 348–351.

Signalment. (n.d.). In The Free Dictionary by Farlex. Retrieved from http://medical-dictionary.thefreedictionary.com/signalment

Stiles, E. K., Palestrini, C., Beauchamp, G., & Frank, D. (2011). Physiological and behavioral effects of dextroamphetamine on Beagle dogs. *Journal of Veterinary Behavior, 6,* 328–336.

Voigt, L. F., Koepsell, T. D., Nelson, J. L., Dugowson, C. E., & Daling, J. R. (1994). Smoking, obesity, alcohol consumption, and the risk of rheumatoid arthritis. *Epidemiology, 5*(5), 525–532.

Ward, D. J. (1971). Arthritis and personality: A controlled study. *The British Medical Journal, 2*(5757), 297–299.

Wernham, B. G. J., Trumpatori, B., Lipsett, J., Davidson, G., Wackerow, P., Thomson, A., & Lascelles, B. D. X. (2011). dose reduction of meloxicam in dogs with osteoarthritis-associated pain and impaired mobility. *Journal of Veterinary Internal Medicine, 25,* 1298–1305.

Wrigglesworth, D. J., Mort, E. S., Upton, S. L., & Miller, A. T. (2011). Accuracy of the use of triaxial accelerometry for measuring daily activity as a predictor of daily maintenance energy requirement in healthy adult Labrador Retrievers. *American Journal of Veterinary Research, 72*(9), 1151–1155.

Yam, P. S., Penpraze, V., Young, D., Todd, M. S., Cloney, A. D., Houston-Callaghan, K. A., & Reilly, J. J. (2011). Validity, practical utility and reliability of Actigraph accelerometry for the measurement of habitual physical activity in dogs. *Journal of Small Animal Practice, 52,* 86–92.

## APPENDIX A

Session: ○ Day - 10 ○ Day 0 ○ Day 14

(Study ID)

***ACVS-Outcome Measures Project***
***Brief Pain Inventory***

Today's Date: ☐☐ / ☐☐ / ☐☐
Month Day Year

**Description of Pain:**

Rate your dog's pain.

1. Fill in the oval next to the one number that best describes the pain at its **worst** in the last 10 days.
   ○0 ○1 ○2 ○3 ○4 ○5 ○6 ○7 ○8 ○9 ○10
   No Pain Extreme Pain

2. Fill in the oval next to the one number that best describes the pain at its **least** in the last 10 days.
   ○0 ○1 ○2 ○3 ○4 ○5 ○6 ○7 ○8 ○9 ○10
   No Pain Extreme Pain

3. Fill in the oval next to the one number that best describes the pain at its **average** in the last 10 days.
   ○0 ○1 ○2 ○3 ○4 ○5 ○6 ○7 ○8 ○9 ○10
   No Pain Extreme Pain

4. Fill in the oval next to the one number that best describes the pain as it is **right now**.
   ○0 ○1 ○2 ○3 ○4 ○5 ○6 ○7 ○8 ○9 ○10
   No Pain Extreme Pain

**Description of Function:**

Fill in the oval next to the one number that best describes **how during the 10 days pain has interfered with** your dog's:

5. **General Activity**
   ○0 ○1 ○2 ○3 ○4 ○5 ○6 ○7 ○8 ○9 ○10
   Does not Interfere Completely Interferes

6. **Enjoyment of Life**
   ○0 ○1 ○2 ○3 ○4 ○5 ○6 ○7 ○8 ○9 ○10
   Does not Interfere Completely Interferes

7. **Ability to Rise to Standing From Lying Down**
   ○0 ○1 ○2 ○3 ○4 ○5 ○6 ○7 ○8 ○9 ○10
   Does not Interfere Completely Interferes

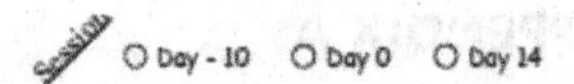


(Study ID)

## *ACVS-Outcome Measures Project*
## *Brief Pain Inventory (Cont)*

**8. Ability to Walk**

O 0 O 1 O 2 O 3 O 4 O 5 O 6 O 7 O 8 O 9 O 10

Does not Interfere … Completely Interferes

**9. Ability to Run**

O 0 O 1 O 2 O 3 O 4 O 5 O 6 O 7 O 8 O 9 O 10

Does not Interfere … Completely Interferes

**10. Ability to Climb Stairs, Curbs, Doorsteps, etc.**

O 0 O 1 O 2 O 3 O 4 O 5 O 6 O 7 O 8 O 9 O 10

Does not Interfere … Completely Interferes

**Overall Impression:**

11. Fill in the oval next to the one number that best describes your dog's overall quality of life over the last 10 days.

O Poor O Fair O Good O Very Good O Excellent

Session: O Day - 10 O Day 0 O Day 14

(Study ID)

# ACVS-Outcome Measures Project
# Caregiver Level of Interaction

**Today's Date:** __ / __ / __
Month Day Year

For each of the following questions, please fill in the oval next to the one number that best describes your interactions with your dog over the last 10 days.

1. Which best describes the overall lifestyle of your dog?

   O Predominantly Indoors O Equally Indoors & Outdoors O Predominantly Outdoors

2. How many days out of the last seven were you in the company of your dog?

   O 0 O 1 O 2 O 3 O 4 O 5 O 6 O 7

3. On average, how many hours in the day were you awake in the company of your dog?

   O None O Less than 1 hour O 1-3 hours O 4-6 hours O 7-10 hours O More than 10 hours

4. On average, how many hours in the day were you actively interacting with you dog (walking, playing with, petting, feeding, grooming, medicating etc.)?

   O None O Less than 1 hour O 1-2 hours O 3-4 hours O 5-6 hours O More than 6 hours

5. During your last 7 nights of sleep, how many nights did your dog sleep in the same room with you?

   O 0 O 1 O 2 O 3 O 4 O 5 O 6 O 7

## ACVS-Outcome Measures Project
## ACVS Questionnaire

**Description of Function:**

Please indicate how much of a problem each of the following activities has been for your dog over the **past 7 days.** Please select **one** answer for each question below.

7. **Rising to standing** after lying down for at least 15 minutes?

O No problems O Mild problems O Moderate problems O Severe problems O Extreme problems

8. **Jumping up** (as in getting into the car or onto the bed) ?

O No problems O Mild problems O Moderate problems O Severe problems O Extreme problems

9. **Jumping down** (as in getting out of the car or off of the bed) ?

O No problems O Mild problems O Moderate problems O Severe problems O Extreme problems

10. **Climbing up** (as in stairs, ramps or curbs) ?

O No problems O Mild problems O Moderate problems O Severe problems O Extreme problems

11. **Climbing down** (as in stairs, ramps or curbs) ?

O No problems O Mild problems O Moderate problems O Severe problems O Extreme problems

12. Posturing (getting into the position) to urinate or defecate?

O No problems O Mild problems O Moderate problems O Severe problems O Extreme problems

# APPENDIX B

## One Factor ANOVA (Activity Level by Breed)

### Summary of Fit

| | |
|---|---|
| Rsquare | 0.261406 |
| Adj Rsquare | 0.261362 |
| Root Mean Square Error | 902.2093 |
| Mean of Response | 209.4537 |
| Observations (or Sum Wgts) | 371520 |

### Analysis of Variance

| Source | DF | Sum of Squares | Mean Square | F Ratio | Prob > F |
|---|---|---|---|---|---|
| Breed | 22 | 1.0702e+11 | 4.8647e+9 | 5976.439 | <.0001* |
| Error | 371497 | 3.0239e+11 | 813981.56 | | |
| C. Total | 371519 | 4.0942e+11 | | | |

### Means Comparisons—Connecting Letters Report

| Level | | | | | | | | | | | Mean |
|---|---|---|---|---|---|---|---|---|---|---|---|
| BOXE | A | | | | | | | | | | 3681.4579 |
| DALM | | B | | | | | | | | | 194.3290 |
| ESSP | | B | C | D | | | | | | | 159.9182 |
| MIXB | | B | C | | | | | | | | 154.1952 |
| OESD | | | C | D | E | F | | | | | 125.6666 |
| GERM | | | | D | E | | | | | | 124.4372 |
| COTH | | | C | D | E | F | G | | | | 118.4844 |
| LABR | | | | D | E | F | | | | | 113.4329 |
| WEIM | | | C | D | E | F | G | H | I | | 112.4932 |
| BULL | | | | D | E | F | G | | | | 111.2703 |
| COSP | | | | D | E | F | G | H | | | 110.0604 |
| ROTW | | | | D | E | F | G | H | | | 105.9450 |
| BEAG | | | | D | E | F | G | H | I | | 102.4956 |
| GLDR | | | | | | F | G | H | I | J | 77.6934 |
| SSPD | | | | | | F | G | H | I | J | 74.6891 |
| AKIA | | | | | E | F | G | H | I | J | 74.5462 |
| DOBE | | | | | E | F | G | H | I | J | 70.1497 |
| HUSK | | | | | | F | G | H | I | J | 67.1443 |
| NELH | | | | | | | G | H | I | J | 59.8074 |
| MASF | | | | | | | | H | I | J | 47.3062 |
| IWFH | | | | | E | F | G | H | I | J | 43.7156 |
| GSMD | | | | | | | | | I | J | 36.0923 |
| BULM | | | | | | | | | | J | 20.8777 |

*Note:* Levels not connected by same letter are significantly different.

## Three Factor ANOVA (Activity Level by Age, Weight and Treatment)

**Analysis of Variance**

| Source | DF | Sum of Squares | Mean Square | F Ratio |
|---|---|---|---|---|
| Model | 11 | 4975324914 | 452302265 | 413.9034 |
| Error | 370068 | 4.044e+11 | 1092772.5 | Prob > F |
| C. Total | 370079 | 4.0938e+11 | | <.0001* |

**Effect Tests**

| Source | Nparm | DF | Sum of Squares | F Ratio | Prob > F |
|---|---|---|---|---|---|
| Treatment | 2 | 2 | 781732512 | 357.6831 | <.0001* |
| AgeD | 1 | 1 | 357240907 | 326.9124 | <.0001* |
| Treatment*AgeD | 2 | 2 | 481491487 | 220.3073 | <.0001* |
| Weight | 1 | 1 | 199662450 | 182.7118 | <.0001* |
| Treatment*Weight | 2 | 2 | 621300165 | 284.2770 | <.0001* |
| AgeD*Weight | 1 | 1 | 154492358 | 141.3765 | <.0001* |
| Treatment*AgeD*Weight | 2 | 2 | 507383797 | 232.1544 | <.0001* |

## Multi Factor ANOVA without Interaction

**Summary of Fit**

| | |
|---|---|
| Rsquare | 0.275468 |
| Rsquare Adj | 0.275401 |
| Root Mean Square Error | 895.2875 |
| Mean of Response | 210.0414 |
| Observation (or Sum Wgts) | 370080 |

**Analysis of Variance**

| Source | DF | Sum of Squares | Mean Square | F Ratio |
|---|---|---|---|---|
| Model | 34 | 1.1277e+11 | 3.3168e+9 | 4137.979 |
| Error | 370045 | 2.9661e+11 | 801539.77 | **Prob > F** |
| C. Total | 370079 | 4.0938e+11 | | <.0001* |

**Parameter Estimates**

| Term | Estimate | Std Error | t Ratio | Prob>\|t\| |
|---|---|---|---|---|
| Intercept | 376.10829 | 9.79857 | 38.38 | <.0001* |
| Treatment[Normal–No] | –126.4952 | 5.634129 | –22.45 | <.0001* |
| Treatment[Yes–Normal] | –110.5724 | 5.495441 | –20.12 | <.0001* |
| Age Group[2–1] | 85.950166 | 6.758527 | 12.72 | <.0001* |
| Age Group[3–2] | –120.6198 | 6.821818 | –17.68 | <.0001* |
| Age Group[4–3] | –23.55457 | 6.05916 | –3.89 | 0.0001* |
| Age Group[5–4] | –58.86602 | 4.871359 | –12.08 | <.0001* |
| Age Group[6–5] | 0.251903 | 8.929011 | 0.03 | 0.9775 |
| SexD[0] | –19.93413 | 1.864915 | –10.69 | <.0001* |
| Weight Group[3–2] | –2.003418 | 7.402983 | –0.27 | 0.7867 |
| Weight Group[4–3] | 128.01285 | 4.680898 | 27.35 | <.0001* |
| Weight Group[5–4] | –125.3011 | 6.752638 | –18.56 | <.0001* |
| Weight Group[6–5] | –41.91263 | 8.461228 | –4.95 | <.0001* |
| Breed[AKIA] | –172.3703 | 13.90485 | –12.40 | <.0001* |
| Breed[BEAG] | –97.4341 | 15.85761 | –6.14 | <.0001* |
| Breed[BOXE] | 3336.1641 | 10.54147 | 316.48 | <.0001* |
| Breed[BULL] | –272.3137 | 11.62624 | –23.42 | <.0001* |
| Breed[BULM] | –347.6137 | 14.56024 | –23.87 | <.0001* |
| Breed[COSP] | –93.22451 | 12.02054 | –7.76 | <.0001* |
| Breed[COTH] | –122.6615 | 15.32197 | –8.01 | <.0001* |
| Breed[DALM] | –190.4761 | 14.00975 | –13.60 | <.0001* |
| Breed[DOBE] | –90.61996 | 14.90055 | –6.08 | <.0001* |
| Breed[ESSP] | –155.0982 | 14.80274 | –10.48 | <.0001* |
| Breed[GERM] | –184.8862 | 7.323063 | –25.25 | <.0001* |
| Breed[GLDR] | –271.7774 | 10.86715 | –25.01 | <.0001* |
| Breed[GSMD] | –128.0755 | 15.64131 | –8.19 | <.0001* |
| Breed[HUSK] | –139.904 | 13.84016 | –10.11 | <.0001* |
| Breed[IWFH] | 0.4829783 | 24.27831 | 0.02 | 0.9841 |
| Breed[LABR] | –148.7832 | 4.262575 | –34.90 | <.0001* |
| Breed[MASF] | –316.5039 | 15.11844 | –20.93 | <.0001* |
| Breed[MIXB] | –85.15796 | 4.191752 | –20.32 | <.0001* |
| Breed[NELH] | –59.34823 | 16.00559 | –3.71 | 0.0002* |
| Breed[OESD] | –169.6185 | 10.22211 | –16.59 | <.0001* |
| Breed[ROTW] | –187.9609 | 10.64088 | –17.66 | <.0001* |
| Breed[SSPD] | –45.86104 | 13.92874 | –3.29 | 0.0010* |

**Effect Tests**

| Source | Nparm | DF | Sum of Squares | F Ratio | Prob > F |
|---|---|---|---|---|---|
| Treatment | 2 | 2 | 4067990661 | 2537.610 | <.0001* |
| Age Group | 5 | 5 | 798490034 | 199.2390 | <.0001* |
| SexD | 1 | 1 | 91580150.1 | 114.2553 | <.0001* |
| Weight Group | 4 | 4 | 828503944 | 258.4101 | <.0001* |
| Breed | 22 | 22 | 9.0916e+10 | 5155.762 | <.0001* |

# APPENDIX C

## Nominal Logistic Fit for Treatment

*Converged in Gradient, 27 Iterations*

### Whole Model Test

| Model | -LogLikelihood | DF | ChiSquare | Prob>ChiSq |
|---|---|---|---|---|
| Difference | 136508.82 | 80 | 273017.6 | <.0001* |
| Full | 247896.00 | | | |
| Reduced | 384404.82 | | | |

| | |
|---|---|
| RSquare (U) | 0.3551 |
| AICc | 495968 |
| BIC | 496920 |
| Observations (or Sum Wgts) | 370080 |

| Measure | Training | Definition |
|---|---|---|
| Entropy RSquare | 0.3551 | 1-Loglike(model)/Loglike(0) |
| Generalized RSquare | 0.5965 | (1-(L(0)/L(model))^(2/n))/(1-L(0)^(2/n)) |
| Mean -Log p | 0.6698 | $\Sigma$ -Log(ρ[j])/n |
| RMSE | 0.4847 | $\sqrt{\Sigma(y[j]-\rho[j])^2/n}$ |
| Mean Abs Dev | 0.4326 | $\Sigma$ \|y[j]−ρ[j]\|/n |
| Misclassification Rate | 0.3385 | $\Sigma$ (ρ[j] ≠ ρMax)/n |
| N | 370080 | n |

### Lack of Fit

| Source | DF | -LogLikelihood | ChiSquare |
|---|---|---|---|
| Lack of Fit | 330 | 204203.34 | 408406.7 |
| Saturated | 410 | 43692.66 | Prob>ChiSq |
| Fitted | 80 | 247896.00 | <.0001* |

### Parameter Estimates

| Term | Estimate | Std Error | ChiSquare | Prob>ChiSq |
|---|---|---|---|---|
| Intercept | −11.241388 | 0.157242 | 5111.0 | <.0001* |
| AgeD | 0.82462908 | 0.0192412 | 1836.8 | <.0001* |
| Weight | 0.34058791 | 0.0051064 | 4448.7 | <.0001* |
| AgeD*Weight | −0.0279494 | 0.0006372 | 1923.8 | <.0001* |

*(continued)*

**Parameter Estimates (continued)**

| Term | | Estimate | Std Error | ChiSquare | Prob>ChiSq |
|---|---|---|---|---|---|
| BPI_General[1–0] | | 14.2149264 | 0.2091869 | 4617.6 | <.0001* |
| BPI_General[2–1] | | –5.4094084 | 0.1683778 | 1032.1 | <.0001* |
| BPI_General[3–2] | | 4.66798085 | 0.1552914 | 903.57 | <.0001* |
| BPI_General[4–3] | | –9.917596 | 0.2162026 | 2104.2 | <.0001* |
| BPI_General[5–4] | | 5.53953156 | 0.2453266 | 509.87 | <.0001* |
| BPI_General[6–5] | | 7.31901038 | 0.2542336 | 828.78 | <.0001* |
| BPI_General[7–6] | | –5.3249153 | 0.2474599 | 463.04 | <.0001* |
| BPI_General[8–7] | | –18.437033 | 0.7371365 | 625.58 | <.0001* |
| BPI_General[9–8] | Biased | 47.740201 | 6793557.5 | 0.00 | 1.0000 |
| BPI_General[10–9] | Biased | 1.9287e–11 | 895896.03 | 0.00 | 1.0000 |
| AgeD*BPI_General[1–0] | | –1.3509019 | 0.025124 | 2891.1 | <.0001* |
| AgeD*BPI_General[2–1] | | 0.74896633 | 0.0198464 | 1424.2 | <.0001* |
| AgeD*BPI_General[3–2] | | –0.3882128 | 0.019148 | 411.05 | <.0001* |
| AgeD*BPI_General[4–3] | | 1.49959617 | 0.0276098 | 2950.0 | <.0001* |
| AgeD*BPI_General[5–4] | | –1.0947104 | 0.0314851 | 1208.9 | <.0001* |
| AgeD*BPI_General[6–5] | | –0.6002606 | 0.0315639 | 361.66 | <.0001* |
| AgeD*BPI_General[7–6] | | 0.27875446 | 0.028514 | 95.57 | <.0001* |
| AgeD*BPI_General[8–7] | | 1.61575439 | 0.0795776 | 412.26 | <.0001* |
| AgeD*BPI_General[9–8] | | –1.5336147 | 687550.03 | 0.00 | 1.0000 |
| AgeD*BPI_General[10–9] | Zeroed | 0 | 0 | . | . |
| Weight*BPI_General[1–0] | | –0.44954 | 0.0069321 | 4205.5 | <.0001* |
| Weight*BPI_General[2–1] | | 0.15308104 | 0.0055406 | 763.35 | <.0001* |
| Weight*BPI_General[3–2] | | –0.1487069 | 0.0050584 | 864.26 | <.0001* |
| Weight*BPI_General[4–3] | | 0.31135666 | 0.0062456 | 2485.2 | <.0001* |
| Weight*BPI_General[5–4] | | –0.0577057 | 0.0069633 | 68.68 | <.0001* |
| Weight*BPI_General[6–5] | | –0.2779082 | 0.0074731 | 1383.0 | <.0001* |
| Weight*BPI_General[7–6] | | 0.10070558 | 0.0068212 | 217.96 | <.0001* |
| Weight*BPI_General[8–7] | | 0.43961917 | 0.0231071 | 361.96 | <.0001* |
| Weight*BPI_General[9–8] | | –0.4114894 | 136443.71 | 0.00 | 1.0000 |
| Weight*BPI_General[10–9] | Zeroed | 0 | 0 | . | . |
| AgeD*Weight*BPI_General[1–0] | | 0.04555151 | 0.0008432 | 2918.2 | <.0001* |
| AgeD*Weight*BPI_General[2–1] | | –0.0221313 | 0.0006688 | 1095.0 | <.0001* |
| AgeD*Weight*BPI_General[3–2] | | 0.01497802 | 0.0006463 | 537.03 | <.0001* |
| AgeD*Weight*BPI_General[4–3] | | –0.046974 | 0.0008382 | 3140.3 | <.0001* |
| AgeD*Weight*BPI_General[5–4] | | 0.01938522 | 0.0009422 | 423.30 | <.0001* |
| AgeD*Weight*BPI_General[6–5] | | 0.02619686 | 0.0009852 | 707.01 | <.0001* |
| AgeD*Weight*BPI_General[7–6] | | –0.0043307 | 0.0008624 | 25.21 | <.0001* |
| AgeD*Weight*BPI_General[8–7] | | –0.0325277 | 0.0025381 | 164.24 | <.0001* |
| AgeD*Weight*BPI_General[9–8] | | 0.02780136 | 13948.371 | 0.00 | 1.0000 |
| AgeD*Weight*BPI_General[10–9] | Biased | 0 | 0 | . | . |
| Intercept | | –4.2740171 | 0.1159704 | 1358.2 | <.0001* |
| AgeD | | 0.72084635 | 0.0148759 | 2348.1 | <.0001* |

*(continued)*

**Parameter Estimates (continued)**

| Term | | Estimate | Std Error | ChiSquare | Prob>ChiSq |
|---|---|---|---|---|---|
| Weight | | 0.29492958 | 0.0045777 | 4151.0 | <.0001* |
| AgeD*Weight | | –0.0434454 | 0.0006029 | 5193.5 | <.0001* |
| BPI_General[1–0] | | 8.28494974 | 0.3080747 | 723.22 | <.0001* |
| BPI_General[2–1] | | –6.2465471 | 0.4252201 | 215.80 | <.0001* |
| BPI_General[3–2] | | 4.52707426 | 0.6308998 | 51.49 | <.0001* |
| BPI_General[4–3] | Unstable | –34.115907 | 108524.65 | 0.00 | 0.9997 |
| BPI_General[5–4] | | 3.13271536 | 158352.96 | 0.00 | 1.0000 |
| BPI_General[6–5] | | 3.40845345 | 175247.33 | 0.00 | 1.0000 |
| BPI_General[7–6] | | –2.4549382 | 173923.06 | 0.00 | 1.0000 |
| BPI_General[8–7] | | –11.488563 | 357258.65 | 0.00 | 1.0000 |
| BPI_General[9–8] | Unstable | 39.226779 | 9613723.5 | 0.00 | 1.0000 |
| BPI_General[10–9] | Biased | –1.362e–10 | 1267016 | 0.00 | 1.0000 |
| AgeD*BPI_General[1–0] | | –1.2886682 | 0.0342867 | 1412.6 | <.0001* |
| AgeD*BPI_General[2–1] | | 0.17722955 | 0.051533 | 11.83 | 0.0006* |
| AgeD*BPI_General[3–2] | | –0.7779877 | 0.0806757 | 93.00 | <.0001* |
| AgeD*BPI_General[4–3] | | 1.89506088 | 13093.623 | 0.00 | 0.9999 |
| AgeD*BPI_General[5–4] | | –0.6170901 | 19772.064 | 0.00 | 1.0000 |
| AgeD*BPI_General[6–5] | | –0.2824207 | 21883.188 | 0.00 | 1.0000 |
| AgeD*BPI_General[7–6] | | 0.1229802 | 20094.465 | 0.00 | 1.0000 |
| AgeD*BPI_General[8–7] | | 1.04260809 | 40576.57 | 0.00 | 1.0000 |
| AgeD*BPI_General[9–8] | | –0.9925584 | 973135.89 | 0.00 | 1.0000 |
| AgeD*BPI_General[10–9] | Zeroed | 0 | 0 | . | . |
| Weight*BPI_General[1–0] | | –0.5525306 | 0.0122174 | 2045.3 | <.0001* |
| Weight*BPI_General[2–1] | | 0.19641661 | 0.014802 | 176.08 | <.0001* |
| Weight*BPI_General[3–2] | | –0.199918 | 0.0189467 | 111.34 | <.0001* |
| Weight*BPI_General[4–3] | | 0.37359747 | 2889.0456 | 0.00 | 0.9999 |
| Weight*BPI_General[5–4] | | –0.0413858 | 4472.3644 | 0.00 | 1.0000 |
| Weight*BPI_General[6–5] | | –0.1320772 | 5106.9058 | 0.00 | 1.0000 |
| Weight*BPI_General[7–6] | | 0.0437475 | 4814.5771 | 0.00 | 1.0000 |
| Weight*BPI_General[8–7] | | 0.29838198 | 10677.018 | 0.00 | 1.0000 |
| Weight*BPI_General[9–8] | | –0.2811615 | 193237.27 | 0.00 | 1.0000 |
| Weight*BPI_General[10–9] | Biased | 0 | 0 | . | . |
| AgeD*Weight*BPI_General[1–0] | | 0.07065677 | 0.0013767 | 2634.0 | <.0001* |
| AgeD*Weight*BPI_General[2–1] | | –0.0107726 | 0.001767 | 37.17 | <.0001* |
| AgeD*Weight*BPI_General[3–2] | | 0.03218546 | 0.0024304 | 175.37 | <.0001* |
| AgeD*Weight*BPI_General[4–3] | | –0.0685305 | 385.25296 | 0.00 | 0.9999 |
| AgeD*Weight*BPI_General[5–4] | | 0.01180926 | 597.30268 | 0.00 | 1.0000 |
| AgeD*Weight*BPI_General[6–5] | | 0.01259103 | 679.45239 | 0.00 | 1.0000 |
| AgeD*Weight*BPI_General[7–6] | | –0.0016142 | 611.15337 | 0.00 | 1.0000 |
| AgeD*Weight*BPI_General[8–7] | | –0.0244859 | 1259.7574 | 0.00 | 1.0000 |
| AgeD*Weight*BPI_General[9–8] | | 0.02160603 | 19763.55 | 0.00 | 1.0000 |
| AgeD*Weight*BPI_General[10–9] | Zeroed | 0 | 0 | . | . |

*For log odds of No/Yes, Normal/Yes*

**Effect Likelihood Ratio Tests**

| Source | Nparm | DF | L-R ChiSquare | Prob>ChiSq |
|---|---|---|---|---|
| AgeD | 2 | 2 | 2712.27212 | <.0001* |
| Weight | 2 | 2 | 5866.87751 | <.0001* |
| AgeD*Weight | 2 | 2 | 6582.54016 | <.0001* |
| BPI_General | 20 | 20 | 9924.27635 | <.0001* |
| AgeD*BPI_General | 20 | 18 | 9238.68627 | <.0001* |
| Weight*BPI_General | 20 | 18 | 11212.0615 | <.0001* |
| AgeD*Weight*BPI_General | 20 | 18 | 12125.1954 | <.0001* |

# APPENDIX D

## Least Squares Regression for Activity Level

*Response AvgOfActivity*

**Summary of Fit**

| | |
|---|---|
| RSquare | 0.69049 |
| RSquare Adj | 0.690358 |
| Root Mean Square Error | 593.1639 |
| Mean of Response | 213.3081 |
| Observations (or Sum Wgts) | 360000 |

**Analysis of Variance**

| Source | DF | Sum of Squares | Mean Square | F Ratio |
|---|---|---|---|---|
| Model | 153 | 2.8245e+11 | 1.8461e+9 | 5246.958 |
| Error | 359846 | 1.2661e+11 | 351843.43 | Prob > F |
| C. Total | 359999 | 4.0906e+11 | | <.0001* |

**Parameter Estimates**

| Term | Estimate | Std Error | t Ratio | Prob>\|t\| |
|---|---|---|---|---|
| Intercept | 135156.82 | 5005.304 | 27.00 | <.0001* |
| Time | 0.0010908 | 3.964e–5 | 27.52 | <.0001* |
| Treatment[No] | 510.16676 | 3.998157 | 127.60 | <.0001* |
| Treatment[Normal] | –521.3443 | 6.439823 | –80.96 | <.0001* |

*(continued)*

**Parameter Estimates (continued)**

| Term | Estimate | Std Error | t Ratio | Prob>\|t\| |
|---|---|---|---|---|
| AgeD | −23.69099 | 0.856791 | −27.65 | <.0001* |
| SexD[0] | −97.80759 | 1.842838 | −53.07 | <.0001* |
| Weight | −1.279587 | 0.256592 | −4.99 | <.0001* |
| Breed[AKIA] | −27.80241 | 24.92747 | −1.12 | 0.2647 |
| Breed[BEAG] | 1057.2114 | 18.5666 | 56.94 | <.0001* |
| Breed[BOXE] | 5993.0729 | 13.19095 | 454.33 | <.0001* |
| Breed[BULL] | −1104.239 | 14.41696 | −76.59 | <.0001* |
| Breed[BULM] | 228.90629 | 16.29008 | 14.05 | <.0001* |
| Breed[COSP] | −412.822 | 10.11335 | −40.82 | <.0001* |
| Breed[COTH] | −129.2766 | 21.05685 | −6.14 | <.0001* |
| Breed[DALM] | −49.05019 | 15.05896 | −3.26 | 0.0011* |
| Breed[DOBE] | −551.0105 | 16.85387 | −32.69 | <.0001* |
| Breed[ESSP] | −723.054 | 12.88092 | −56.13 | <.0001* |
| Breed[GERM] | −205.1 | 11.02372 | −18.61 | <.0001* |
| Breed[GLDR] | −898.7163 | 9.248697 | −97.17 | <.0001* |
| Breed[GSMD] | 263.54305 | 15.19536 | 17.34 | <.0001* |
| Breed[HUSK] | −65.31997 | 14.73917 | −4.43 | <.0001* |
| Breed[IWFH] | −422.2026 | 34.45078 | −12.26 | <.0001* |
| Breed[LABR] | −334.2661 | 4.868256 | −68.66 | <.0001* |
| Breed[MASF] | −1191.199 | 17.34432 | −68.68 | <.0001* |
| Breed[MIXB] | −329.0445 | 4.209729 | −78.16 | <.0001* |
| Breed[NELH] | −831.5574 | 19.36797 | −42.93 | <.0001* |
| Breed[OESD] | 229.13229 | 11.9801 | 19.13 | <.0001* |
| Breed[ROTW] | −238.7083 | 11.69143 | −20.42 | <.0001* |
| Breed[SSPD] | −212.3764 | 18.15105 | −11.70 | <.0001* |
| BPI_PainWst[1–0] | −450.1489 | 12.56922 | −35.81 | <.0001* |
| BPI_PainWst[2–1] | −252.5286 | 9.776927 | −25.83 | <.0001* |
| BPI_PainWst[3–2] | 273.14894 | 9.181727 | 29.75 | <.0001* |
| BPI_PainWst[4–3] | −28.23764 | 6.486766 | −4.35 | <.0001* |
| BPI_PainWst[5–4] | 336.19695 | 7.406067 | 45.39 | <.0001* |
| BPI_PainWst[6–5] | −298.7649 | 7.259921 | −41.15 | <.0001* |
| BPI_PainWst[7–6] | −493.0149 | 6.749522 | −73.04 | <.0001* |
| BPI_PainWst[8–7] | 585.05109 | 7.523688 | 77.76 | <.0001* |
| BPI_PainWst[9–8] | −815.3079 | 14.3438 | −56.84 | <.0001* |
| BPI_PainWst[10–9] | 834.16498 | 31.6666 | 26.34 | <.0001* |
| BPI_PainLst[1–0] | 45.579018 | 5.283085 | 8.63 | <.0001* |
| BPI_PainLst[2–1] | −43.86924 | 6.102914 | −7.19 | <.0001* |
| BPI_PainLst[3–2] | −51.23109 | 8.10592 | −6.32 | <.0001* |
| BPI_PainLst[4–3] | 191.52009 | 8.855961 | 21.63 | <.0001* |
| BPI_PainLst[5–4] | −301.6556 | 12.88523 | −23.41 | <.0001* |
| BPI_PainLst[6–5] | 172.73062 | 25.05463 | 6.89 | <.0001* |

*(continued)*

**Parameter Estimates (continued)**

| Term | Estimate | Std Error | t Ratio | Prob>\|t\| |
|---|---|---|---|---|
| BPI_PainLst[7–6] | −1531.946 | 56.98255 | −26.88 | <.0001* |
| BPI_PainLst[8–7] | 6069.1449 | 57.20494 | 106.09 | <.0001* |
| BPI_General[1–0] | 493.02934 | 7.188381 | 68.59 | <.0001* |
| BPI_General[2–1] | −666.8271 | 7.552484 | −88.29 | <.0001* |
| BPI_General[3–2] | 17.120791 | 6.802297 | 2.52 | 0.0118* |
| BPI_General[4–3] | 257.30446 | 7.851418 | 32.77 | <.0001* |
| BPI_General[5–4] | 117.9061 | 7.958756 | 14.81 | <.0001* |
| BPI_General[6–5] | −367.0629 | 10.54371 | −34.81 | <.0001* |
| BPI_General[7–6] | 181.51899 | 10.99418 | 16.51 | <.0001* |
| BPI_General[8–7] | 304.09686 | 13.26337 | 22.93 | <.0001* |
| BPI_General[9–8] | −912.2096 | 21.82206 | −41.80 | <.0001* |
| BPI_General[10–9] | −1363.098 | 41.8507 | −32.57 | <.0001* |
| BPI_Life[1–0] | −355.808 | 7.028391 | −50.62 | <.0001* |
| BPI_Life[2–1] | 420.33321 | 5.855803 | 71.78 | <.0001* |
| BPI_Life[3–2] | −520.5862 | 6.275568 | −82.95 | <.0001* |
| BPI_Life[4–3] | 733.33178 | 8.468047 | 86.60 | <.0001* |
| BPI_Life[5–4] | −383.9416 | 12.1509 | −31.60 | <.0001* |
| BPI_Life[6–5] | 233.0212 | 11.92163 | 19.55 | <.0001* |
| BPI_Life[7–6] | −190.0438 | 13.50075 | −14.08 | <.0001* |
| BPI_Life[8–7] | −174.6814 | 15.88 | −11.00 | <.0001* |
| BPI_Life[9–8] | 1857.8594 | 41.21186 | 45.08 | <.0001* |
| BPI_Rise[1–0] | 226.47447 | 9.738803 | 23.25 | <.0001* |
| BPI_Rise[2–1] | −26.06753 | 7.845699 | −3.32 | 0.0009* |
| BPI_Rise[3–2] | 99.323823 | 6.246335 | 15.90 | <.0001* |
| BPI_Rise[4–3] | 395.92811 | 7.271005 | 54.45 | <.0001* |
| BPI_Rise[5–4] | −1040.267 | 7.393602 | −140.7 | <.0001* |
| BPI_Rise[6–5] | 248.73075 | 8.523633 | 29.18 | <.0001* |
| BPI_Rise[7–6] | −108.3741 | 10.80647 | −10.03 | <.0001* |
| BPI_Rise[8–7] | 242.85394 | 9.528686 | 25.49 | <.0001* |
| BPI_Rise[9–8] | 264.70029 | 14.29821 | 18.51 | <.0001* |
| BPI_Rise[10–9] | 187.99116 | 19.43166 | 9.67 | <.0001* |
| BPI_Walking[1–0] | 590.89145 | 7.124123 | 82.94 | <.0001* |
| BPI_Walking[2–1] | −204.3278 | 5.789229 | −35.29 | <.0001* |
| BPI_Walking[3–2] | −396.1804 | 6.336951 | −62.52 | <.0001* |
| BPI_Walking[4–3] | 116.40291 | 8.731668 | 13.33 | <.0001* |
| BPI_Walking[5–4] | 288.07138 | 12.71866 | 22.65 | <.0001* |
| BPI_Walking[6–5] | 229.87493 | 13.64997 | 16.84 | <.0001* |
| BPI_Walking[7–6] | −654.8327 | 12.86142 | −50.91 | <.0001* |
| BPI_Walking[8–7] | 8.837146 | 20.6188 | 0.43 | 0.6682 |
| BPI_Walking[9–8] | 1633.4677 | 52.5087 | 31.11 | <.0001* |
| BPI_Run[1–0] | −421.6262 | 8.651548 | −48.73 | <.0001* |

*(continued)*

**Parameter Estimates (continued)**

| Term | Estimate | Std Error | t Ratio | Prob>\|t\| |
|---|---|---|---|---|
| BPI_Run[2–1] | 229.6734 | 7.103353 | 32.33 | <.0001* |
| BPI_Run[3–2] | –158.6128 | 6.984806 | –22.71 | <.0001* |
| BPI_Run[4–3] | 162.04011 | 7.010634 | 23.11 | <.0001* |
| BPI_Run[5–4] | 470.07851 | 8.419369 | 55.83 | <.0001* |
| BPI_Run[6–5] | –1061.138 | 12.05696 | –88.01 | <.0001* |
| BPI_Run[7–6] | 58.601255 | 12.40552 | 4.72 | <.0001* |
| BPI_Run[8–7] | 574.17775 | 9.624408 | 59.66 | <.0001* |
| BPI_Run[9–8] | –114.8531 | 10.68223 | –10.75 | <.0001* |
| BPI_Run[10–9] | 229.83526 | 16.3669 | 14.04 | <.0001* |
| BPI_Climb[1–0] | –437.267 | 9.266196 | –47.19 | <.0001* |
| BPI_Climb[2–1] | 664.75739 | 8.277699 | 80.31 | <.0001* |
| BPI_Climb[3–2] | –207.0222 | 6.165939 | –33.58 | <.0001* |
| BPI_Climb[4–3] | –29.84349 | 7.401107 | –4.03 | <.0001* |
| BPI_Climb[5–4] | –157.0956 | 8.850792 | –17.75 | <.0001* |
| BPI_Climb[6–5] | 352.32253 | 9.481218 | 37.16 | <.0001* |
| BPI_Climb[7–6] | 159.00191 | 10.31857 | 15.41 | <.0001* |
| BPI_Climb[8–7] | –48.90406 | 12.98304 | –3.77 | 0.0002* |
| BPI_Climb[9–8] | –146.9909 | 14.54393 | –10.11 | <.0001* |
| BPI_Climb[10–9] | 282.05975 | 18.90556 | 14.92 | <.0001* |
| BPI_Overall[2–1] | –186.6596 | 35.9659 | –5.19 | <.0001* |
| BPI_Overall[3–2] | 381.98057 | 10.50683 | 36.36 | <.0001* |
| BPI_Overall[4–3] | 105.84538 | 5.579858 | 18.97 | <.0001* |
| BPI_Overall[5–4] | –84.82056 | 5.561343 | –15.25 | <.0001* |
| LOI_Lifestyle[1–0] | 2735.3337 | 18.66885 | 146.52 | <.0001* |
| LOI_Lifestyle[2–1] | –153.4594 | 5.342185 | –28.73 | <.0001* |
| LOI_Lifestyle[3–2] | –16.70426 | 19.44996 | –0.86 | 0.3904 |
| LOI_Lifestyle[4–3] | 666.03011 | 27.00375 | 24.66 | <.0001* |
| LOI_Days[4–2] | 5542.6704 | 28.23726 | 196.29 | <.0001* |
| LOI_Days[5–4] | –5025.766 | 23.28511 | –215.8 | <.0001* |
| LOI_Days[6–5] | 177.62865 | 18.31712 | 9.70 | <.0001* |
| LOI_Days[7–6] | –73.02285 | 14.20685 | –5.14 | <.0001* |
| LOI_Hours[3–2] | –144.6158 | 9.659341 | –14.97 | <.0001* |
| LOI_Hours[4–3] | 70.666404 | 3.961272 | 17.84 | <.0001* |
| LOI_Hours[5–4] | –218.3111 | 6.001743 | –36.37 | <.0001* |
| LOI_Interact[2–1] | 301.92464 | 8.945314 | 33.75 | <.0001* |
| LOI_Interact[3–2] | –127.1849 | 4.048175 | –31.42 | <.0001* |
| LOI_Interact[4–3] | –78.9393 | 6.270432 | –12.59 | <.0001* |
| LOI_Interact[5–4] | 141.00424 | 7.482617 | 18.84 | <.0001* |
| LOI_Nights[1–0] | –134.6668 | 11.22263 | –12.00 | <.0001* |
| LOI_Nights[2–1] | 723.84103 | 16.49727 | 43.88 | <.0001* |
| LOI_Nights[3–2] | –1470.142 | 18.04004 | –81.49 | <.0001* |

*(continued)*

**Parameter Estimates (continued)**

| Term | Estimate | Std Error | t Ratio | Prob>\|t\| |
|---|---|---|---|---|
| LOI_Nights[4–3] | 384.00384 | 14.02289 | 27.38 | <.0001* |
| LOI_Nights[5–4] | 766.67832 | 13.78838 | 55.60 | <.0001* |
| LOI_Nights[6–5] | –69.73754 | 14.44565 | –4.83 | <.0001* |
| LOI_Nights[7–6] | –182.1538 | 10.71873 | –16.99 | <.0001* |
| _7Func Rise[1–0] | 77.131676 | 7.806885 | 9.88 | <.0001* |
| _7Func Rise[2–1] | 136.87978 | 5.719343 | 23.93 | <.0001* |
| _7Func Rise[3–2] | 235.96515 | 10.07891 | 23.41 | <.0001* |
| _7Func Rise[4–3] | 5576.0676 | 85.12307 | 65.51 | <.0001* |
| 11FuncClDo[1–0] | –68.17276 | 6.991189 | –9.75 | <.0001* |
| 11FuncClDo[2–1] | 63.420849 | 5.980522 | 10.60 | <.0001* |
| 11FuncClDo[3–2] | –88.96714 | 11.8896 | –7.48 | <.0001* |
| 11FuncClDo[4–3] | –764.2317 | 30.65502 | –24.93 | <.0001* |
| 10FuncClUp[1–0] | 149.80257 | 7.466963 | 20.06 | <.0001* |
| 10FuncClUp[2–1] | –333.0446 | 6.054379 | –55.01 | <.0001* |
| 10FuncClUp[3–2] | 433.9337 | 10.15312 | 42.74 | <.0001* |
| 10FuncClUp[4–3] | –4969.522 | 70.31338 | –70.68 | <.0001* |
| 8FuncJuUp[1–0] | –250.6216 | 7.477251 | –33.52 | <.0001* |
| 8FuncJuUp[2–1] | 180.39342 | 5.624834 | 32.07 | <.0001* |
| 8FuncJuUp[3–2] | –184.0659 | 7.373677 | –24.96 | <.0001* |
| 8FuncJuUp[4–3] | –512.6554 | 11.98984 | –42.76 | <.0001* |
| 9FuncJuDo[1–0] | –173.6149 | 6.416716 | –27.06 | <.0001* |
| 9FuncJuDo[2–1] | 124.07597 | 4.961373 | 25.01 | <.0001* |
| 9FuncJuDo[3–2] | 19.83004 | 10.26014 | 1.93 | 0.0533 |
| 9FuncJuDo[4–3] | 1956.2944 | 34.28091 | 57.07 | <.0001* |
| 12FuncPost[1–0] | –107.6925 | 4.930343 | –21.84 | <.0001* |
| 12FuncPost[2–1] | –307.9655 | 6.416623 | –47.99 | <.0001* |
| 12FuncPost[3–2] | 812.16771 | 15.89011 | 51.11 | <.0001* |

**Effect Tests**

| Source | Nparm | DF | Sum of Squares | F Ratio | Prob > F |
|---|---|---|---|---|---|
| Time | 1 | 1 | 266448775 | 757.2936 | <.0001* |
| Treatment | 2 | 2 | 7762158668 | 11030.70 | <.0001* |
| AgeD | 1 | 1 | 269008394 | 764.5685 | <.0001* |
| SexD | 1 | 1 | 991106622 | 2816.897 | <.0001* |
| Weight | 1 | 1 | 8749916.85 | 24.8688 | <.0001* |
| Breed | 22 | 22 | 1.0508e+11 | 13575.39 | <.0001* |
| BPI_PainWst | 10 | 10 | 4978472754 | 1414.968 | <.0001* |
| BPI_PainLst | 8 | 8 | 4178060714 | 1484.347 | <.0001* |
| BPI_General | 10 | 10 | 5418816333 | 1540.122 | <.0001* |
| BPI_Life | 9 | 9 | 6306647261 | 1991.621 | <.0001* |

*(continued)*

**Effect Tests (continued)**

| Source | Nparm | DF | Sum of Squares | F Ratio | Prob > F |
|---|---|---|---|---|---|
| BPI_Rise | 10 | 10 | 8746871940 | 2486.013 | <.0001* |
| BPI_Walking | 9 | 9 | 6010408510 | 1898.069 | <.0001* |
| BPI_Run | 10 | 10 | 6848088556 | 1946.345 | <.0001* |
| BPI_Climb | 10 | 10 | 3362856639 | 955.7821 | <.0001* |
| BPI_Overall | 4 | 4 | 762535418 | 541.8145 | <.0001* |
| LOI_Lifestyle | 4 | 4 | 9306528648 | 6612.692 | <.0001* |
| LOI_Days | 4 | 4 | 2.4074e+10 | 17105.68 | <.0001* |
| LOI_Hours | 3 | 3 | 617471660 | 584.9872 | <.0001* |
| LOI_Interact | 4 | 4 | 937919338 | 666.4323 | <.0001* |
| LOI_Nights | 7 | 7 | 3474796215 | 1410.853 | <.0001* |
| _7Func Rise | 4 | 4 | 2126284548 | 1510.817 | <.0001* |
| 11FuncClDo | 4 | 4 | 328839303 | 233.6546 | <.0001* |
| 10FuncClUp | 4 | 4 | 3723516406 | 2645.720 | <.0001* |
| 8FuncJuUp | 4 | 4 | 1997551968 | 1419.347 | <.0001* |
| 9FuncJuDo | 4 | 4 | 1720578839 | 1222.546 | <.0001* |
| 12FuncPost | 3 | 3 | 2482143544 | 2351.561 | <.0001* |

## APPENDIX E
## Decision Tree for Treatment

**Fit Details**

| Measure | Training | Definition |
|---|---|---|
| Entropy RSquare | 0.9055 | 1-Loglike(model)/Loglike(0) |
| Generalized RSquare | 0.9691 | (1-(L(0)/L(model))^(2/n))/(1-L(0)^(2/n)) |
| Mean -Log p | 0.0984 | ∑ -Log(ρ[j])/n |
| RMSE | 0.1802 | $\sqrt{\sum(y[j]-\rho[j])^2/n}$ |
| Mean Abs Dev | 0.0650 | ∑ \|y[j]-ρ[j]\|/n |
| Misclassification Rate | 0.0465 | ∑ (ρ[j]≠ρMax)/n |
| N | 371520 | n |

**Confusion Matrix**

Actual / Predicted

| Training | No | Normal | Yes |
|---|---|---|---|
| No | 142560 | 0 | 12960 |
| Normal | 0 | 67680 | 0 |
| Yes | 4320 | 0 | 144000 |

**Crossvalidation**

| k-fold | -2LogLike | RSquare |
|---|---|---|
| 5 Folded | 73058.9913 | 0.9056 |
| Overall | 73051.3749 | 0.9055 |

**Column Contributions**

| Term | Number of Splits | G^2 | Portion |
|---|---|---|---|
| BPI_PainWst | 4 | 274794.554 | 0.3922 |
| Breed | 5 | 92674.3049 | 0.1323 |
| BPI_Rise | 4 | 67190.2954 | 0.0959 |
| BPI_Climb | 3 | 60779.6459 | 0.0867 |
| BPI_Run | 3 | 47202.5132 | 0.0674 |
| BPI_PainNow | 1 | 32921.5354 | 0.0470 |
| BPI_Life | 3 | 24913.8456 | 0.0356 |
| BPI_General | 2 | 18263.2348 | 0.0261 |
| LOI_Interact | 1 | 16837.8975 | 0.0240 |
| Weight | 1 | 16258.8393 | 0.0232 |
| BPI_Walking | 1 | 15331.3571 | 0.0219 |
| LOI_Nights | 1 | 14400.4898 | 0.0206 |
| 11FuncClDo | 1 | 10354.8024 | 0.0148 |
| 8FuncJuUp | 1 | 8764.31614 | 0.0125 |

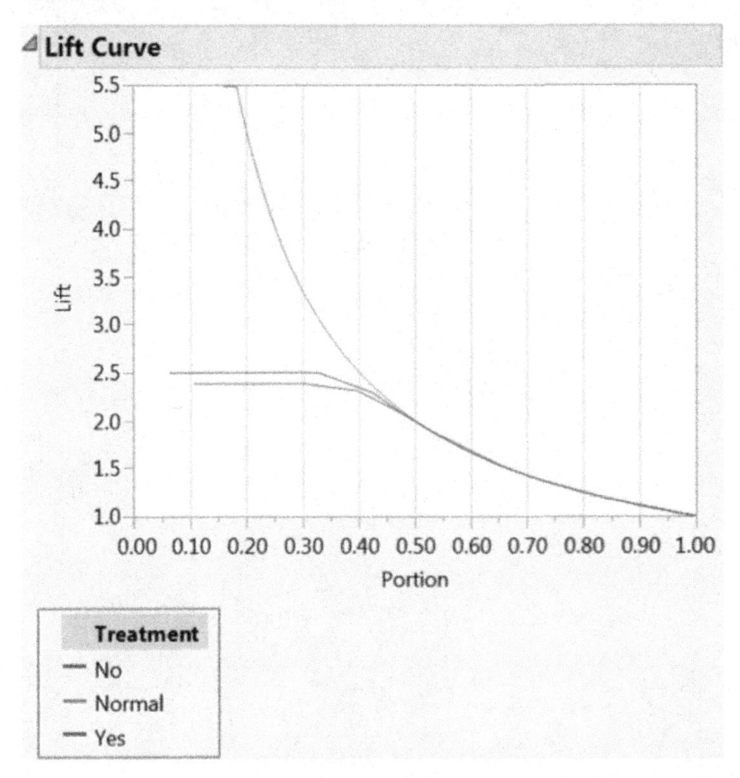

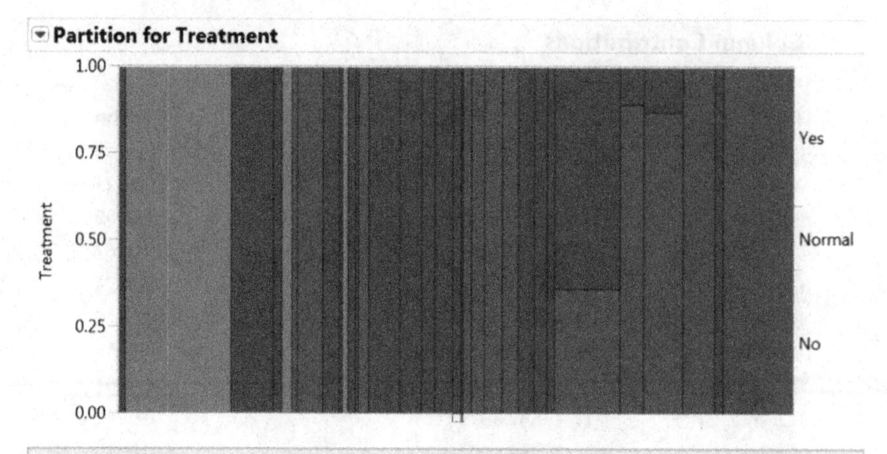
Partition for Treatment
1.00
0.75
0.50
0.25
0.00
Treatment
Yes
Normal
No

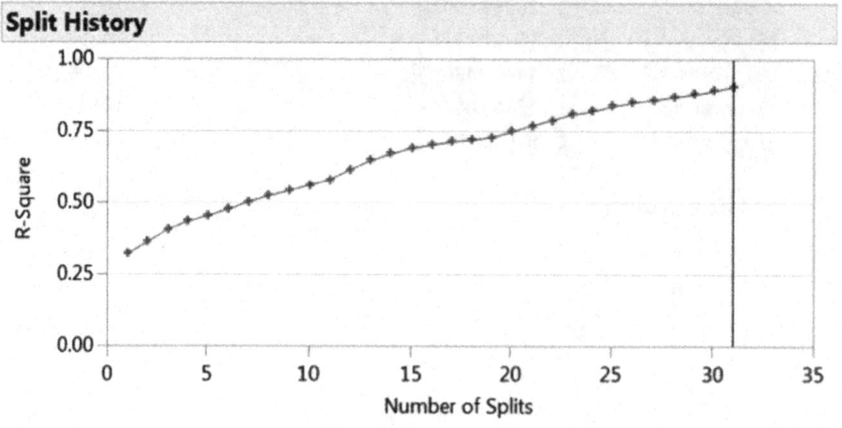
Split History
1.00
0.75
0.50
0.25
0.00
R-Square
0
5
10
15
20
25
30
35
Number of Splits

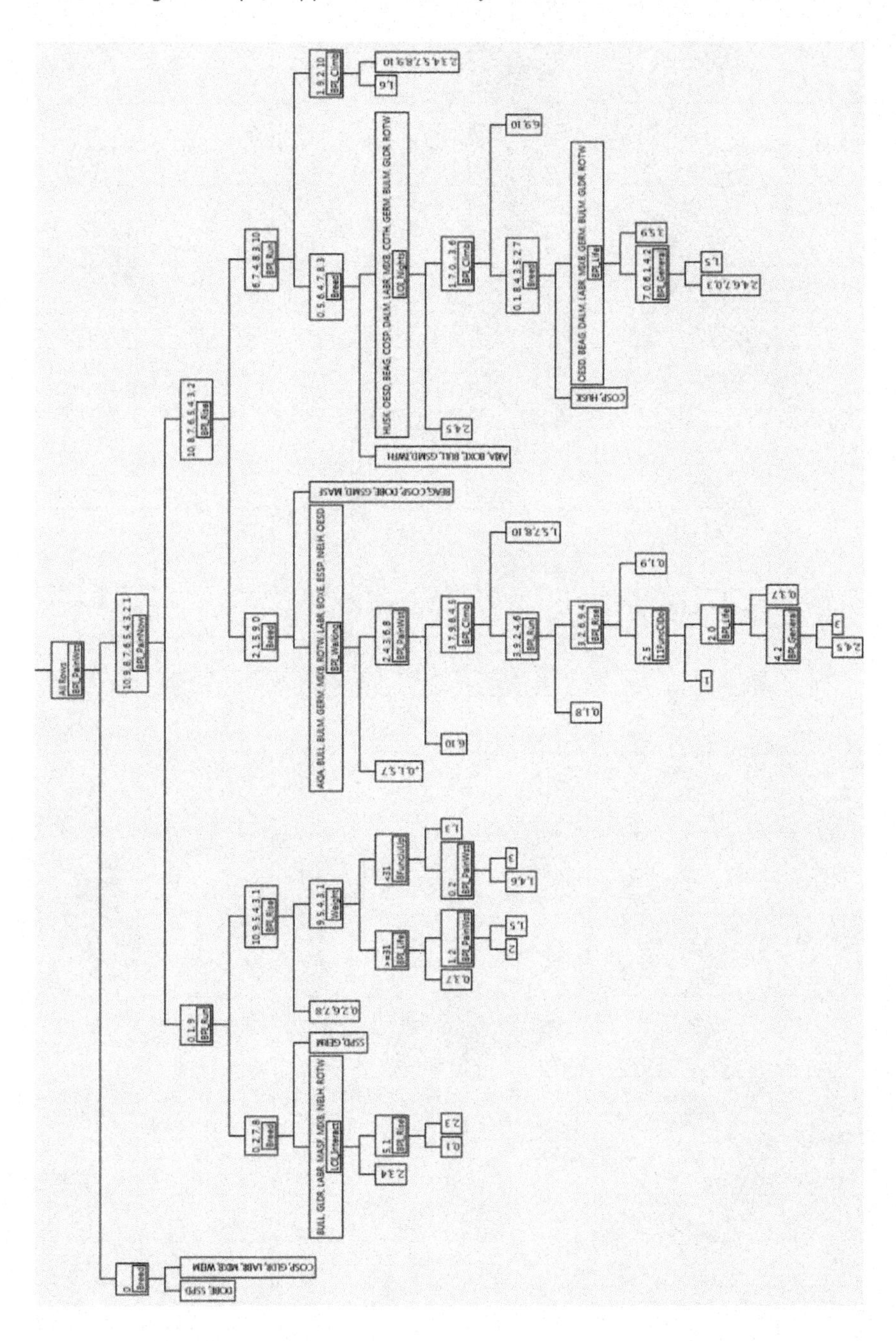

CHAPTER 9

# DATA MINING TECHNIQUES FOR INFORMATION ASSURANCE AND DATA INTEGRITY ON THE CLOUD

**Alla Kammerdiner**
***New Mexico State University***

## ABSTRACT

Security threats to information systems become more persistent and sophisticated. This trend is reflected by increasing costs of cybercrimes for business and government alike. Information assurance is an integral part of an organization's cyber security. Growing dependence of organizations on cloud storage and computing services raises new threats to data integrity. In the cloud, your data are distributed over networked storage, communication, and computing devices provided by a third party. Although cloud providers may have a strong incentive for keeping the integrity of your data in order to maintain trust, an organization should be able to independently monitor the integrity of its data. In this chapter, a new approach to information assurance and data integrity on the cloud is proposed. The cloud resources and the resources of the organization are modeled together as a network. This joint network is monitored via the information exchange between trusted nodes. Statistical

*Contemporary Perspectives in Data Mining, Volume 2*, pages 177–190

inference is combined with network analyses to support the decision-making related to information assurance. In many application domains, including energy and finance, an organization's data are examples of big data. Therefore, the chapter discusses the potential for the proposed approach to be applied on such data and be utilized for applications other than cloud computing.

## INTRODUCTION

The economic costs of cybercrime in the United States are about $70 billion to $140 billion per year according to early estimates (Lewis & Baker, 2013). Conservative estimates of the annual global losses from illegal internet activities are between $80 billion and $400 billion, although other sources estimate global losses from malicious cyber activity to be up to $1 trillion annually (Lewis & Baker, 2013). With criminals steadily moving their fraudulent activities from the real world to the internet, internet crime continues to rise with a reported 8.3% increase in unverified losses reported to IC3 during 2012 (Federal Bureau of Investigation, 2013).

At the same time, the society's reliance on the internet continues to grow rapidly in the United States, with ever decreasing cost and increasing availability of telecommunications (Sanou, 2013). From online banking and shopping to online education and telemedicine, the cyber world is an important part of our daily lives. Even critical infrastructure, the backbone of our nation's economy, security, and health, is becoming dependent on the internet raising concerns about its cyber security (The White House, 2013). A part of this trend is a rapid emergence of cloud computing.

Cloud computing outsources data storage and data processing to the providers adept at building and managing large datacenters at low cost. The economies of scale increase returns for cloud providers and lower computing costs for cloud users (Chen, Paxson, & Katz, 2010). Similar to more traditional outsourcing, cloud computing offers economic and operational advantages, including an ability to focus on core activities, cost and efficiency savings, reduced overhead, operational control, and flexibility. Other key advantages that are unique to cloud computing include greater accessibility and less dependency on local physical resources.

Security is widely recognized to be the major barrier for the widespread adoption of cloud computing (Armbrust et al., 2009; Chen et al., 2010; European Network and Information Security Agency, 2009; Mell & Grance, 2009; Shankland, 2009). Of course, some security issues associated with cloud computing are due to its inherent dependence on internet. Undoubtedly, there are two security aspects that are "new and fundamental to cloud computing: the complexities of multi-party trust considerations, and the ensuing need for mutual auditability" (Chen et al., 2010, p. 1).

Cloud computing obscures some information from the organization and leads to uncertainty about the data. In the age of information technology, information assurance and data integrity play a crucial role in the success of a business. Reportedly, information theft remains the highest external cost of cybercrime, accounting for 40–44% of external consequences of cybercrimes in 2010–2013 (Ponemon Institute, 2013). In the cloud environment, the data becomes a new asset that needs to be protected. Although some may argue that this function should be entrusted to cloud providers, the author believes that the organization should play an active role in monitoring the integrity of their critical data.

Data mining and data science can provide effective solutions to the security and trust challenges posed by emergence of cloud computing technologies. Because of its multiparty structure, cloud computing innately obscures some information about other parties. New methods for knowledge discovery can be developed to answer key question about security and trustworthiness of hidden structures from incomplete data. This chapter illustrates this idea by presenting recently proposed method for data mining on network for monitoring data integrity.

The remainder of the chapter is organized as follows. In the next section, we present a motivation for our network-based data mining approach through defining cloud computing, describing the key challenges to data integrity posed by cloud computing, and reviewing the current solutions. Then we introduce a network-based methodology for detection of compromised connections and illustrate it with a simple example. Next, we discuss the contribution of the proposed approach to data mining and its relation to traditional data mining techniques such as decision trees and neural networks. Finally, we present our conclusions.

## CLOUD COMPUTING AND DATA INTEGRITY ON THE CLOUD

The notion of cloud computing has been evolving along with emergence and spread of cloud technologies. An early definition dating back to the beginning of 2009 states that cloud computing includes application software delivered as services over the internet, along with the hardware and systems software in the data centers to support these services (Armbrust et al., 2009). An alternative definition given by the European Network and Information Security Agency describes cloud computing as an on-demand service model for provision of information technology that is “often based on virtualization and distributed computing” (Catteddu & Hogben, 2009; p. 14). The U.S. National Institute of Standards and Technology (NIST) produced a more complete definition. There, cloud computing is defined as “a model for enabling ubiquitous, convenient, on-demand network access to a shared pool of configurable computing resources (e.g., networks, servers, storage, applications, and services) that can be rapidly provisioned and released with minimal management effort or service provider interaction" (Mell & Grance, 2011, p. 2).

The key features of cloud computing include on-demand self-service, broad network access, resource pooling, rapid elasticity or expansion, and measured service (Mell & Grance, 2011). Software as a service (SaaS), platform as a service (PaaS), and infrastructure as a service (IaaS) are three fundamental service models included under the umbrella of cloud computing. In the context of the utilization and delivery of cloud services, cloud computing offers four deployment models, namely public cloud (e.g., Google Docs), community cloud (e.g., wikis, volunteer computing), private cloud, and hybrid cloud. By design, cloud computing architecture has a complex multiplayer nature that involves the five major actors (or roles): cloud user/consumer, cloud provider, cloud auditor, cloud broker, and cloud carrier (Liu et al., 2011).

Nine of the greatest vulnerabilities within cloud computing are identified in the most recent report, which surveyed professional opinion of industry experts. Data breaches are determined to be the highest-ranked critical threat to cloud computing in terms of the severity, followed by data loss (Los, Gray, Shackleford, & Sullivan, 2013). Data breaches represent a situation when sensitive internal data of the organization become available to its competitors. Although data breach predates cloud computing, the advent of the cloud introduces significant new avenues of attack (Los et al., 2013). For example, a virtual machine can potentially use the information on side channel timing to extract private cryptographic keys of other virtual machines on the same server (Zhang, Juels, Reiter, & Ristenpart, 2012). Moreover, a data breach can be easily perpetrated through a flaw in a cloud service database shared by multiple cloud consumers. In such a case, vulnerability in one client's application can be exploited to the advantage of another party allowing an attacker to access not only that client's data, but potentially every other client's data. High relevance and substantial implications of data breach have made this threat raise rapidly in the rankings from the number five in 2010 to number one in 2013 (Los et al., 2013).

Data loss is the second most severe threat in cloud computing. Data is an important asset not only for businesses but also for individuals. Many factors can lead to data loss, such as a malicious attacker, an accidental deletion, a loss of an encryption key, or a physical catastrophe. The prevention of data loss is not the sole responsibility of the cloud provider. The cloud consumers can also be liable. For example, after uploading encrypted data onto the cloud, a consumer must preserve the key or risk loss of the cloud data. One example of severe implications of data loss is noncompliance as a result of a failure to submit the organization's data for a mandatory audit. Not surprisingly, data loss has moved from the fifth[1] to the second place among the nine top critical threats to cloud computing from 2010 to 2013 (Los et al., 2013).

A recent systematic review (Hashizume, Rosado, Fernández-Medina, & Fernandez, 2013) of the literature on security issues in cloud computing indicates that existing studies focus on understanding vulnerabilities and developing solutions mostly for those threats that are inherited by cloud computing from other technologies it uses. In particular, data security is often viewed as a function of cloud providers with cloud consumers assuming responsibility for proper data backup strategies (Hashizume et al., 2013). However, this approach to data integrity does not take into account two innovative security aspects unique to cloud computing: the complexities of multiparty trust considerations and the resulting need for mutual auditability (Chen et al., 2010).

The multiparty nature of cloud computing conceals some connections whose existence may violate the integrity of data or lead to loss of data. The organization's ability to audit its critical data could be used not only for detection of any unauthorized changes or loss of data, but importantly, to find the causes and attribute them to other parties. The organization could use partial knowledge of the connections in a cloud to discover some hidden connections responsible for data breach or data loss. Due to variability of data access on the cloud, this discovery will likely require processing of large and complex datasets.

In what follows, we describe our recently proposed approach (Kammerdiner, 2013; Kammerdiner, in press) for data mining of networked systems that uses partial knowledge of a network to investigate hidden connections. In this approach, the monitored data are compared to themselves at different times. First, a partial network model is constructed to represent a multiparty system of a cloud. The goal is to discover unknown connections that are responsible for undesirable changes to data. The discovery uses the information from the data storage resources on a cloud or in an organization that are involved in the comparison of critical data. To aid in inferences, we propose decoupling the network node representing these storage resources into two separate nodes referred to as a source and a destination. Some clouds can be very large, which requires improved scalability. Consequently, for each party in a cloud system, one should combine as many similar resources together into a single node as possible without sacrificing the validity of a resulting network model.

## NETWORK-BASED METHODOLOGY FOR DETECTION OF COMPROMISED CONNECTIONS

In this section we describe a network-based method for assessing the integrity of data on the cloud. First, we formally define the network model that represents cyber physical connections joining the resources of the organization and its cloud providers. Next, we discuss estimation of the network features via data mining techniques and present an algorithm for testing assumptions about the trustworthiness of the cloud providers.

Let $G = (V, E)$ be a connected graph, where a set $V$ of nodes represents storage and information resources of the organization and its cloud providers, and a set $E$ of edges denotes communication links between various resources. Suppose that the organization can monitor the information flow through the network $G$ via a pair of trusted nodes $s, t \in V$. Specifically, suppose we send the information from $s$ to $t$. Then $s$ becomes a source node and $t$ becomes a sink node of the network on graph $G$.

Next, we describe information about the network—in particular, the communication between nodes and actions on nodes. Let $p_{ij}$ denote a rate at which the information is exchanged between network resources $i$ and $j$. Naturally, $p_{ij} \geq 0$ for any $i, j \in V$. We assume that $p_{ij}$ values are known for some pairs of nodes.

Both types of resources or nodes, namely those belonging to the organization and the ones available via the cloud, could potentially be compromised. Compromised nodes may reduce the value of the data transferred through them in some way. For example, the data might lose some of its value because the compromised node alters these data or discloses the data to cybercriminals or competitors. We model this by introducing a vector-valued function $f$ representing action on nodes. Let $n = |V|$, then $f = (f_1, \cdots, f_n)$, where $f_i = f_i(x)$ denotes action of node $i$ on data $x$. Because sink $s$ and source $t$ are trusted nodes, we have $f_s(x) = x$ and $f_t(x) = x$. For compromised node $c \in V$, by definition $f_c(x) \neq x$ for some $x$. The choice of functions for compromised nodes should be determined by a measurable impact of the possible actions performed on the data by a given compromised resource. When analyzing data integrity, one might be interested in assessing whether the data were altered. In the context, where a compromised node $c$ reverses all data bits, we consider function $f_c(x) = 1 - x$, $x = 0, 1$.

Because the cloud portion of the network, including some connections, is obscured from the organization, some unknown connections via compromised nodes may exist. Therefore, communication uncertainty is another aspect that should be incorporated in the model. To represent possible hidden connections to some cloud resources, let $E^*$ denote unknown edges. Then we can incorporate communication uncertainty into the network model as follows:

- $H_0 : E^* \not\subset E$, meaning that according to our hypothesis, there is no communication via unknown edge
- $H_1 : E^* \subset E$, meaning that under the alternative, communication may involve unknown edges

In addition to communication uncertainty, we consider variability in routing of the information through the network. This means the data can be sent along a number of alternative routes. We model this by introducing randomness in routing from source $s$ to sink $t$ in $G$. Specifically, let $\gamma = \gamma(s,t) = (s, i_1, \cdots, i_{L-2}, t)$ be a random path where $i_k \in V$, $k = 1,\ldots,L-2$.

When we need to assess the integrity of the data, the information is sent from $s$ to $t$ multiple times. During the $j$th transmission of the information $x$ between $s$ and $t$, we get $\gamma_j$, a realization of $\gamma$, and observe $X_j$ on $t$. Let us denote $\gamma_j = (s, i_1^j, \cdots, i_{L-2}^j, t)$ for each $j = 1,\ldots, J$. Then

$$X_j = f_{i_{L-2}^j}\left(\ldots\left(f_{i_1^j}(x)\right)\ldots\right).$$

Given uncertainty in routing from $s$ to $t$, consider a random graph $\Gamma$, which is obtained from $G$ by restricting its vertex and edge set so that $V_\Gamma = V|_{\gamma=\gamma(s,t)}$ and $E_\Gamma = E|_{\gamma=\gamma(s,t)}$. Random graph $\Gamma$ induces a probability distribution $P_\Gamma$. Suppose that for any $e^* \in E^*$ there exists a path $\gamma_0 = \gamma_0(s,t)$ in $\Gamma$ such that $e^* \in \gamma_0(s,t)$. Then, under the hypothesis $H_0$, we get a different graph $\Gamma$ than $\Gamma$ obtained under the alternative $H_1$. Similarly, the induced probability distribution $P_\Gamma$ is different under $H_0$ as compared to $P_\Gamma$ under $H_1$. These differences can be used to differentiate between the alternative $H_0$ and the hypothesis $H_1$.

We formulate a general approach for statistical inference about the presence of compromised connections as the following algorithm:

1. Based on a sample $\overline{X} = (X_1,\ldots,X_n)$ observed on a sink node $t$, estimate empirical distribution $\hat{P}(X|\overline{X})$ of information change $X(s,t) = X$.
2. For the hypothesis $H_0$, estimate distribution $P_\Gamma(X = y|H_0)$ for all $y$, based on the graph $\Gamma$, the node actions $f_i$, and the information exchange frequencies $p_{ij}$.
3. For the alternative $H_1$, estimate distribution $P_\Gamma(X = y|H_1)$ for all $y$, based on $\Gamma$, $f_i$, and $p_{ij}$.
4. Compare the distribution $P_\Gamma(X = y|H_0)$ under the hypothesis $H_0$ to the distribution $P_\Gamma(X = y|H_1)$ under the alternative $H_1$, and determine whether and under what conditions there exist sufficient differences between $P_\Gamma(X = y|H_0)$ and $P_\Gamma(X = y|H_1)$.

5. Apply statistical procedures to assess whether the distribution P $P_\Gamma(X = y|H_0)$ under the hypothesis $H_0$ and the estimated empirical distribution $\hat{P}(X|\bar{X})$ are significantly different.
    - Accept the hypothesis $H_0$, if the distribution $P_\Gamma(X = y|H_0)$ is a good fit for the sample $\bar{X}$.
    - Reject $H_0$ and accept the alternative $H_1$, otherwise.

Below we present a simple example to illustrate application of the introduced network-based methodology for detection of compromised connections.

## Example

Consider a case where a cloud provider monitors two consumers to determine whether data leakage or breach is happening from cloud consumer 1 to consumer 2. Suppose that data breach leads to a verifiable loss of data integrity (e.g., the cloud provider can detect that the consumer's data have been maliciously altered or deleted). We will show how to construct a model and apply our methodology in order to derive statistical procedures for testing whether a hidden connection was used in an attack.

As mentioned earlier, we can construct a network model for this simple system. For the purpose of inference, cloud storage of the critical data belonging to consumer 1 can be first combined and then decoupled to form a source and a sink of a network. In this network, the resources of consumer 1 and consumer 2 are modeled by two separate nodes 1 and 2, respectively. For each consumer, we can aggregate both its local and its cloud resources into a single node.

Based on partial information, known and hypothesized connections among consumer 1 data on the cloud, consumer 1 resources and consumer 2 resources can be defined. In the network model, these connections are represented by the edges. The edges $(s, 1)$ from $s$ to 1 and $(1, t)$ from 1 to $t$ represent data access by consumer 1. These connections are known and do not change the integrity of the data. The edges $(s, 2)$ from $s$ to 2 and $(2, t)$ from 2 to $t$ lead to compromised integrity of consumer 1 data. These edges reflect information about a known vulnerability (e.g., due to multi-tenancy [Hashizume et al., 2013]) that may expose consumer 1 data to consumer 2. The presence of these edges on the network is also known. On the other hand, it is unknown whether consumer 2 can affect the resources of cloud consumer 1 with a purpose of compromising the integrity of consumer 1 data (e.g., as a result of PaaS vulnerability, the security of data can be violated while it is being processed, transferred or stored [Hashizume et al., 2013]). We model these unknown potential connections via

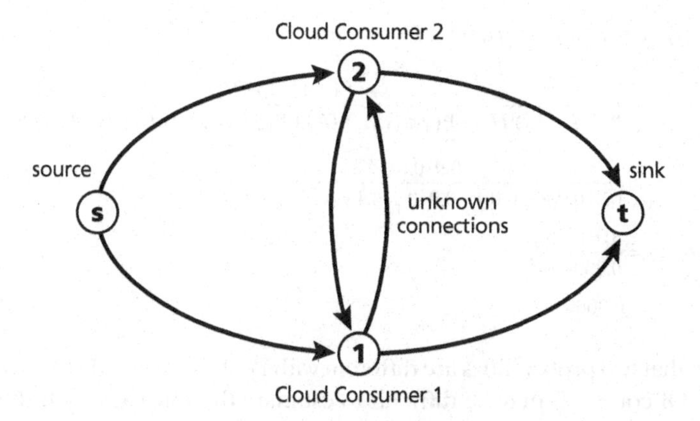


**Figure 9.1** A network model for the case of two cloud consumers monitored via decoupled sink/source.

two hypothesized edges, $(1,2)$ and $(2,1)$. The constructed network model is visualized in Figure 9.1.

Using the network model, we construct the procedure to uncover hidden edges $(1,2)$ and $(2,1)$ from data sample $\bar{X} = (x_1,\ldots,x_n)$ observed on sink $t$. This sample describes changes in the integrity of consumer 1 cloud data. Let $T = T(X) = T(X_1,\ldots,X_n)$ be a random variable representing the number of data with preserved integrity. Since the integrity has binary outcomes, then $T$ is binomial. Hence, empirical distribution $\hat{P}(X|\bar{X})$ of information change $X(s, t) = X$ can be modeled by binomial distribution of $T$. This binomial distribution depends on two parameters, known $n$ and unknown $\mu = \mathrm{P}(X_i = 1)$. Therefore, the procedure to uncover hidden edges can be formulated as inference of $\mu$.

Suppose the estimated rates at which the information is exchanged between resources are $p_{s1} = p_{1t} = 0.9$ and $p_{s2} = p_{2t} = p_{12} = p_{21} = 0.1$. We need to compare the base scenario $H_0$ where edges $(1,2)$ and $(2,1)$ are not present to the alternative scenario $H_1$ where these edges exist. Hence, for $H_0$ and $H_1$ we compute the respective probabilities that the integrity for each compared data $i$ is preserved:

$$
\begin{aligned}
\mathrm{P}_\Gamma(X_i = 1|H_0) &= \mathrm{P}_\Gamma\left((s,1,t) \in \Gamma | H_0\right) \\
&= \frac{\mathrm{P}\left(\gamma \in (s,1,t)|H_0\right)}{\mathrm{P}\left(\gamma \in (s,1,t)|H_0\right) + \mathrm{P}\left(\gamma \in (s,2,t)|H_0\right)} \\
&= \frac{0.9 \cdot 0.9}{0.9 \cdot 0.9 + 0.1 \cdot 0.1} \\
&\approx 0.9878
\end{aligned}
$$

$$\begin{aligned}
P_\Gamma(X_i = 1|H_1) &= P_\Gamma\left((s,1,t) \in \Gamma|H_1\right) \\
&= \frac{P\left(\gamma \in (s,1,t)|H_1\right)}{P\left(\gamma \in (s,1,t)|H_1\right) + P\left(\gamma \in (s,2,t)|H_1\right) + P\left(\gamma \in (s,1,2,t)|H_1\right) + P\left(\gamma \in (s,2,1,t)|H_1\right)} \\
&= \frac{0.9 \cdot 0.9}{0.9 \cdot 0.9 + 0.1 \cdot 0.1 + 0.9 \cdot 0.1 \cdot 0.1 + 0.1 \cdot 0.1 \cdot 0.9} \\
&= \frac{0.81}{0.838} \\
&\approx 0.9666
\end{aligned}$$

Note that two probabilities are different with $P_\Gamma(X_i = 1|H_1) < P_\Gamma(X_i = 1|H_0) = 0.9878$. Of course, it may be difficult to estimate the rates at which the information is exchanged between resources of cloud consumers 1 and 2—that is, $p_{12} = p_{21} = p \geq 0$ is unknown. However, even in the case when $p$ is unknown, the statement $P_\Gamma(X_i = 1|H_1) < P_\Gamma(X_i = 1|H_0) = 0.9878$ remains true, because $p \geq 0$.

Due to this inequality for the probabilities under $H_0$ and $H_1$, we use a one-sided test of the hypothesis regarding the unknown parameter $\mu$ of binomial distribution of $T$. To test $\mu = 0.9878$, we select a significance level $\alpha = 0.01$ and perform the test using the $p$-value method.

Consequently, the procedure for discovering hidden edges can be stated as follows:

1. Compute the $p$-value as the probability $P\left(t \leq T(\bar{X})\right)$, where $t$ is a binomially distributed random variable with parameters $n$ and $\mu = 0.9878$ and $T(\bar{X})$ is the number of data with preserved integrity in a data sample $\bar{X} = (x_1, \ldots, x_n)$.
2. Reject $H_0$ and uncover hidden edges $(1, 2)$ and $(2, 1)$ if $p$-value $\leq \alpha$, otherwise the data do not contradict the known model without these edges.

## DISCUSSION

Data mining is the science and art of data-driven knowledge discovery, which combines approaches from such diverse disciplines as computer science, statistics, and mathematical programming. *Big data* is the term used to describe collections of datasets characterized by high volume, variety, and velocity (Zikopoulos & Eaton, 2011). The overarching goal of the presented methodology is to enable knowledge discovery about networked systems from big data. Therefore, this methodology makes a contribution to the interdisciplinary field of data mining. As illustrated above, our methodology is applicable for extracting useful knowledge about potential vulnerabilities that could exist in multiparty relationships in a cloud computing environment. Undoubtedly, monitoring data integrity on a cloud to uncover hidden multiparty relations not only adds to our capacity to deal with big data but also presents new challenges (Chen & Zhang, 2014).

In many important application domains such as energy and finance, an organization's data are examples of big data. Networked systems often arise when investigating problems in energy (Heussen, Koch, Ulbig, & Andersson, 2012) or finance (Nagurney & Siokos, 1997). Our methodology is constructed in such a way as to enable understanding of complex networked systems with multiparty relationships. Therefore, the presented approach can potentially be applied on big data generated by energy and financial systems, which are becoming increasingly dependent on cyber technologies.

Although our methodology is novel, the network-based approaches are not entirely new to data mining. Some of well-established data mining techniques that use network or graph models include classification trees (in particular, decision trees), regression trees, neural networks, hidden Markov models, Bayesian probabilistic graphical models, and distributed knowledge networks (Ye, 2003). Each of these network-based approaches differs from one another either in the type of a graph or in the way a graph model is used for knowledge discovery.

The presented approach is also unique both in terms of the type of a network graph and in the way a network model is used for extracting new knowledge. For example, a network model utilized in our method and the model in probabilistic graphical networks both have probabilities associated with the edges. Both networks are represented via directed graphs. But, in our approach in the contrast to probabilistic graphical networks, a directed graph does not have to be acyclic. Additionally, our method deals with discovering and analyzing the structure of networked systems rather than complex distributions.

To some degree, the presented network-based approach for data mining of multiparty relations on a cloud is similar to two-class classification. However, instead of classifying observations or data points, our method works by classifying possible edges of the network into two classes, the existing edges and the nonexisting ones. Alternatively, the presented approach could be seen as a new form of outlier detection. Traditionally, outlier detection identifies the data points that deviate markedly from other observations in the sample. In other words, outliers do not fit into a structure defined by most of the data sample. Outliers are routinely removed before performing further analyses of data. Our method identifies "outlier" edges instead of outlier points in the data. In our method, in contrast to traditional outlier detection, outlier edges do not fit into a structure of the network given the data, and so these edges are determined to be nonexistent.

## CONCLUSIONS

We have presented a novel methodology for data mining of multiparty relations on the cloud. The presented models and methods can be used for discovering hidden threats to the integrity of cloud data. The approach can be applied from the point of view of either a cloud provider, or a cloud consumer, and possibly other cloud actors. Our methodology can potentially be extended to other applications where the multiparty networked systems can be studied from big data. We have discussed novel contribution of the proposed approach to data mining and how our approach relates to traditional data mining methods.

## NOTE

1. In the original report/study (Top Threats to Cloud Computing, Version 1.0) issued by CSA in 2010, some cloud security threats were combined resulting in the list of seven threats. In particular, in 2010 data breach and data loss were listed together under data loss or leakage on the fifth out of seven places. In the new 2013 report, the list was split or expanded from the top seven into the top nine threats, with data breach and data loss moving to the first and the second top places, respectively, on the updated nine-item list. Similarly, in 2010 threat #6 was account and service hijacking. In the 2013 version, the sixth threat was split into two separate threats, account hijacking (#3) and denial of service (#5).

## REFERENCES

Armbrust, M., Fox, A., Griffith, R., Joseph, A. D., Katz, R., Konwinski, A., . . . Zaharia, M. (2009). *Above the clouds: A Berkeley view of cloud computing* (Technical Report No. UCB/EECS-2009-28). Berkeley, CA: University of California Berkeley.

Catteddu, D. & Hogben, G. (Eds.). (November 2009). *Cloud computing: Benefits, risks, and recommendations for information security.* Heraklion, Greece: European Network and Information Security Agency (ENISA).

Chen, C. P., & Zhang, C. Y. (2014). Data-intensive applications, challenges, techniques and technologies: A survey on Big Data. *Information Sciences, 275,* 314–347.

Chen, Y., Paxson, V., & Katz, R. H. (2010). *What's new about cloud computing security* (Technical Report No. UCB/EECS-2010-5). Berkley, CA: University of California–Berkley.

European Network and Information Security Agency. (2009, November 20). *Cloud computing risk assessment.* Heraklion, Greece: Author.

Federal Bureau of Investigation. (2013, May 14). IC3 2012 Internet crime report released: More than 280,000 complaints of online criminal activity (national press release). FBI National Press Office. Retrieved from http://www.fbi.gov/news/pressrel/press-releases/ic3-2012-internet-crime-report-released

Hashizume, K., Rosado, D. G., Fernández-Medina, E., & Fernandez, E. B. (2013). An analysis of security issues for cloud computing. *Journal of Internet Services and Applications, 4*(1), 1–13.

Heussen, K., Koch, S., Ulbig, A., & Andersson, G. (2012). Unified system-level modeling of intermittent renewable energy sources and energy storage for power system operation. *IEEE Systems Journal, 6*(1), 140–151.

Kammerdiner, A. (2013, July). *Statistical assessment of cyber threats in cloud computing environment.* Paper presented at the European Conference on Operational Research—EURO 2013, Rome, Italy.

Kammerdiner, A. R. (2014). Statistical techniques for assessing cyberspace security. In C. Vogiatzis, J. L. Walteros, & P. M. Pardalos (Eds.), *Dynamics of information systems* (pp. 161–177). Springer International Publishing.

Lewis, J., & Baker, S. (2013, July). *The economic impact of cybercrime and cyber espionage.* Center for Strategic and International Studies. McAfee, An Intel Company. Retrieved from http://www.mcafee.com/us/resources/reports/rp-economic-impact-cybercrime.pdf

Liu, F., Tong, J., Mao, J., Bohn, R., Messina, J., Badger, L., & Leaf, D. (2011). *NIST cloud computing reference architecture. Recommendations of the National Institute of Standards and Technology.* NIST. Special Publication 500-292. Washington, DC: National Institute of Standards and Technology.

Los, R., Gray, D., Shackleford, D., & Sullivan, B. (2013). *The notorious nine: Cloud computing top threats in 2013.* Seattle, WA: Cloud Security Alliance (CSA), Top Threats Working Group.

Mell, P., & Grance, T. (2009). *Effectively and securely using the cloud computing paradigm.* Gaithersburg, MD: National Institute of Standards and Technology.

Mell, P., & Grance, T. (2011). The NIST Definition of Cloud Computing (NIST Special Publication 800-145). Washington, DC: National Institute of Standards and Technology. Retrieved from http://csrc.nist.gov/publications/nistpubs/800-145/SP800-145.pdf

Nagurney, A., & Siokos, S. (1997). *Financial networks.* Berlin, Germany: Springer.

Ponemon Institute. (2013). Cost of cyber crime study: United States. Retrieved from http://media.scmagazine.com/documents/54/2013_us_ccc_report_final_6-1_13455.pdf

Sanou, B. (2013). ICT facts and figures: The World in 2013. ICT Data and Statistics Division. International Telecommunication Union. Geneva, Switzerland. Retrieved from http://www.itu.int/en/ITU-D/Statistics/Documents/facts/ICTFactsFigures2013-e.pdf

Shankland, S. (2009, October 20). HP's Hurd dings cloud computing, IBM. *CNET News.* Retrieved from http://www.cnet.com/news/hps-hurd-dings-cloud-computing-ibm/

The White House. (2013, February 13). *Executive order: improving critical infrastructure cybersecurity.* The White House. Office of the Press Secretary. Retrieved from http://www.whitehouse.gov/the-press-office/2013/02/12/executive-order-improving-critical-infrastructure-cybersecurity

Ye, N. (Ed.). (2003). *The handbook of data mining* (Vol. 24). Mahwah, NJ: Lawrence Erlbaum Associates.
Zhang, Y., Juels, A., Reiter, M. K., & Ristenpart, T. (2012). Cross-VM side channels and their use to extract private keys. In T. Yu, G. Danezis, & V. Gligor (Eds.), *Proceedings of the 2012 ACM conference on Computer and communications security* (pp. 305–316). New York, NY: Association for Computing Machinery.
Zikopoulos, P., & Eaton, C. (2011). *Understanding big data: Analytics for enterprise class hadoop and streaming data.* New York, NY: McGraw-Hill Osborne Media.

CHAPTER 10

# MULTIVARIATE COPULAS MODEL IN SPATIOTEMPORAL IRREGULAR PATTERN DETECTION IN MOBILITY NETWORK

**Rong Duan and Guang-Qin Ma**
***AT&T Labs Research, Bedminster, NJ***

## ABSTRACT

Characterizing localized mobility network traffic is one of the most challenging tasks in mobility network planning. Network dimensioning does not only need to consider statistically steady-state traffic, but it also needs to take into account the situations when special events happen in some areas, especially when the events incur intensive traffic loads even though the occurrence is rare. This chapter proposes a multivariate copulas concept to identify the areas that have different traffic patterns compared with their neighbors. Multivariate dependence statistic *pseudo* is constructed to measure the spatial relationship among the multiple neighbor time series, which is based on the degree of dependence for multivariate extreme value copula. Weighted dependency, which integrates temporal extreme value features with spatial

*Contemporary Perspectives in Data Mining, Volume 2*, pages 191–213

dependency structure, is established to detect the areas that have irregular temporal traffic patterns. A new spatial neighbor detection procedure is illustrated to obtain the areas that are robust to irregular shapes. A synthetic dataset that simulates the real network traffic is generated to illustrate our procedure and validate the performance.

## INTRODUCTION

With the arrival of the mobile Internet computing era, mobile communications have penetrated in our daily life, and the trend is accelerated by the continued evolution of wireless network technology, the constant enhancement of mobile devices, and the fast development of mobile applications. The rapidly expanded area not only creates opportunities but also brings challenges for wireless service providers. How to optimize the investment and improve customer experience with the tremendous traffic growth is becoming more and more critical to the success of any service provider. Understanding the current network traffic characterization is the first step for business planners to optimize the future investment strategy.

Mobility network traffic and performance are complicated by many dynamic factors and diverse geographical features. In general, most cell sites have day-of-the-week and hour-of-the-day periodic traffic patterns. However, cell sites at different geographical locations present substantially different traffic patterns related to local human flow characteristics. For example, cell sites around museums, amusement parks, and shopping malls have higher traffic load during weekends compared to weekdays; cell sites at airports have higher traffic load than the sites at farm lands; cell sites along highways have similar network traffic patterns with highway traffic, and so on. Even though the traffic patterns are different across different locations in these sites, the periodic patterns for each site are consistent, and the impacted neighbor cells are stable. It is still relatively easy to assign dimension to these types of network traffic. There is another type of location where the volume of network traffic load for the surrounding cell sites and the number of impacted neighbor cell sites are driven by the different events in that area. The network traffic loads are dramatically high when certain events happen. For example, a football stadium will have higher network traffic load and more cell sites involved when it hosts the Super Bowl compared with the regular seasonal games. Large social events attract cell phone users to relatively small areas, which increases the instability of network traffic and downgrades the quality of service. Understanding the traffic characteristics of these areas and identifying the impacted cell sites will help in network capacity planning and enhance the network performance. In this study we focus on detecting the areas that have irregular traffic patterns compared with their neighbors by using BusyHour temporal distribution and dependence among spatial neighbor cell sites.

BusyHour(*BH*) is an important measurement that network engineers use to evaluate and plan network capacity. According to American National Standard for Telecommunication (ATIS, 2015), *BH* is defined as the 60-minute period during which the maximum total traffic load occurs in a given 24-hour period. *BH* is characterized by a block maxima extreme value distribution by definition if the hourly traffic load is an *i.i.d* random variable. To transfer the hourly traffic load to *i.i.d* random variable, periodic patterns have to be removed before constructing the *BH*.

Figures 10.1 and 10.2 illustrate two types hourly network traffic volume. Figure 10.1 represents the site that has consistent patterns. Panel (a) is the

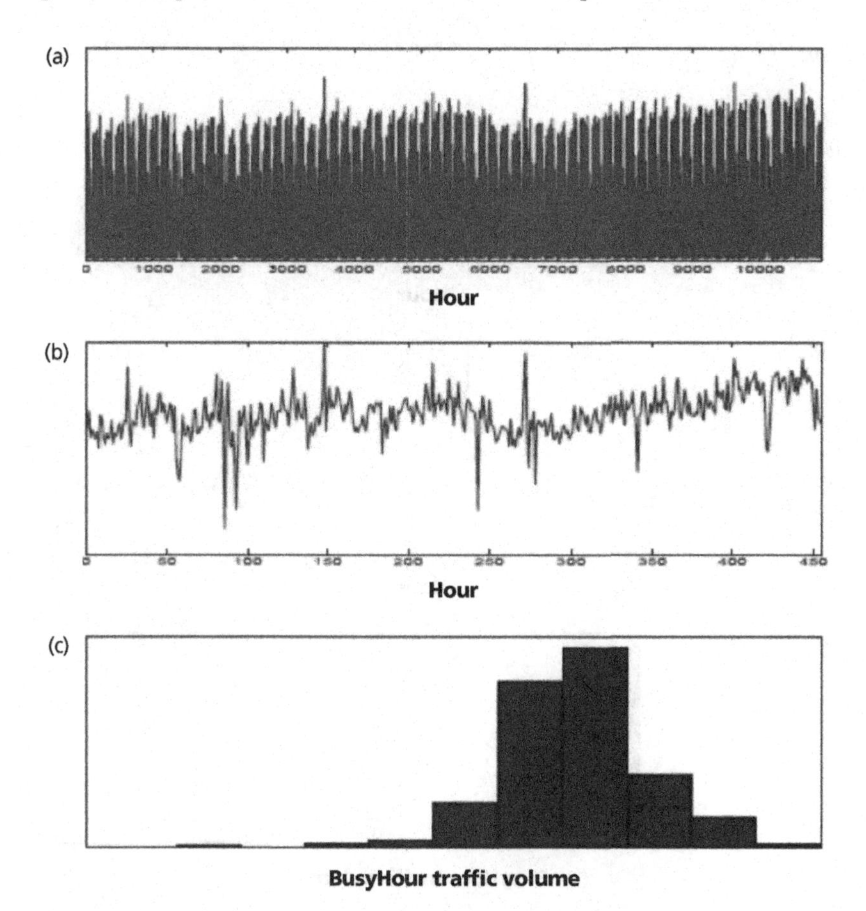


**Figure 10.1** Temporal traffic properties for normal cell site (a) 455 days hourly traffic; (b) 455 days normalized BusyHour traffic without periodic patterns; (c) histogram of normalized BusyHour traffic.

collection of 455 days hourly traffic volume. It shows the daily and hourly periodic patterns: Weekday has higher network traffic load than weekend, daytime has higher load than nighttime, and midnight has the lowest traffic load. Panel (b) is the standardized *BH* for the 455 days after eliminating the daily and hourly periodic patterns. There are many off-the-shelf technologies to remove the periodic patterns. We adopt the median absolute deviation (MAD) method for its simplicity and robustness. Since the scale is not the focus in this study, we normalize *BH*, and its histogram is shown in panel (c).

Figure 10.2 shows the 455-day temporal patterns for a special cell site. There are three significant events that have happened during the 455 days. And the *BH* distribution of this cell site shows a heavy tail in Figure 10.2(c).

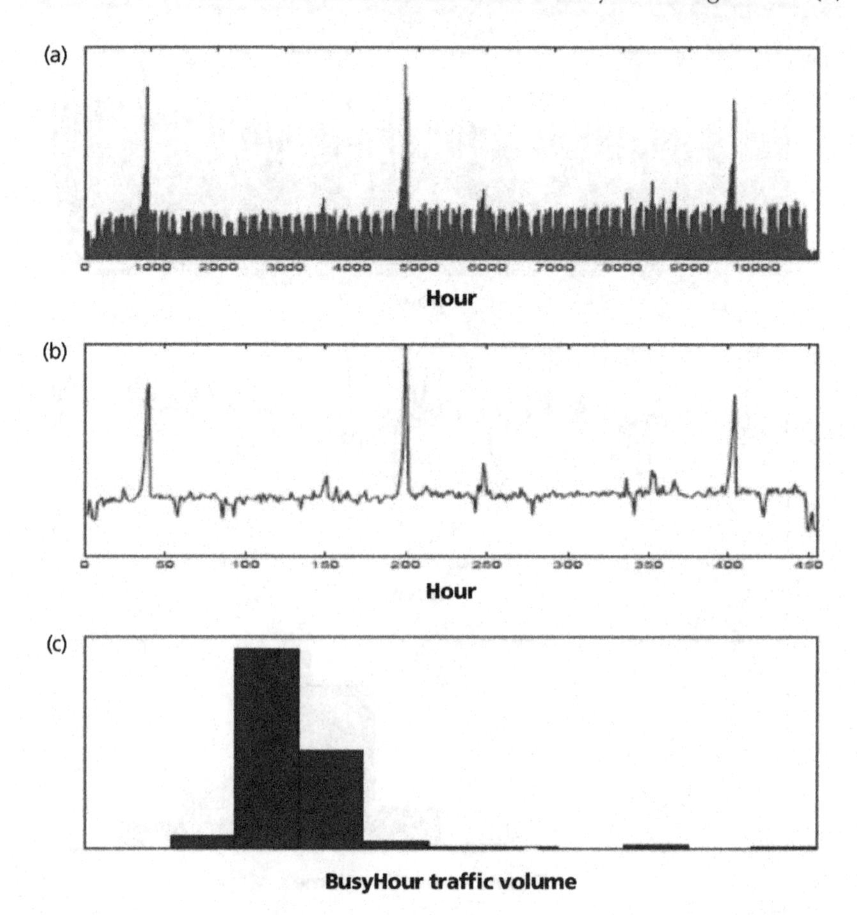


**Figure 10.2** Temporal traffic properties for a Venue site (a) 455 days hourly traffic; (b) 455 days normalized BusyHour traffic without periodic patterns; (c) histogram of normalized BusyHour traffic.

Figures 10.1 and 10.2 illustrate the different temporal patterns between regular cell sites and the sites with special events. The histogram of the special event cell site is more positively skewed than the normal site.

Another challenge of the problem is the dependency among the spatial neighbor. There is not just one cell site that serves an area, especially for areas that have large special events. As shown in Figure 10.3, panel (a) is the relative spatial relations for cell sites. Each point represents one cell site. The cell sites *A, B* and *C* that are marked as plus have some common events, and the sites *D,E, F* that are marked with a triangle are randomly picked. The temporal patterns for these six sites are shown in panel (b). *A, B* and *C* have different temporal patterns from *D, E* and *F*. A, B, and C's traffic is more spiky than D, E, and F. *B* and *C's* patterns are more similar compared to A, since B and C are spatially closer. The neighbor cell sites share similar temporal patterns. The temporal patterns are not spatially *i.i.d.*; they are correlated to each other. It can be seen that the temporal pattern for the cell sites co-occurred if they are close enough. The objective for this

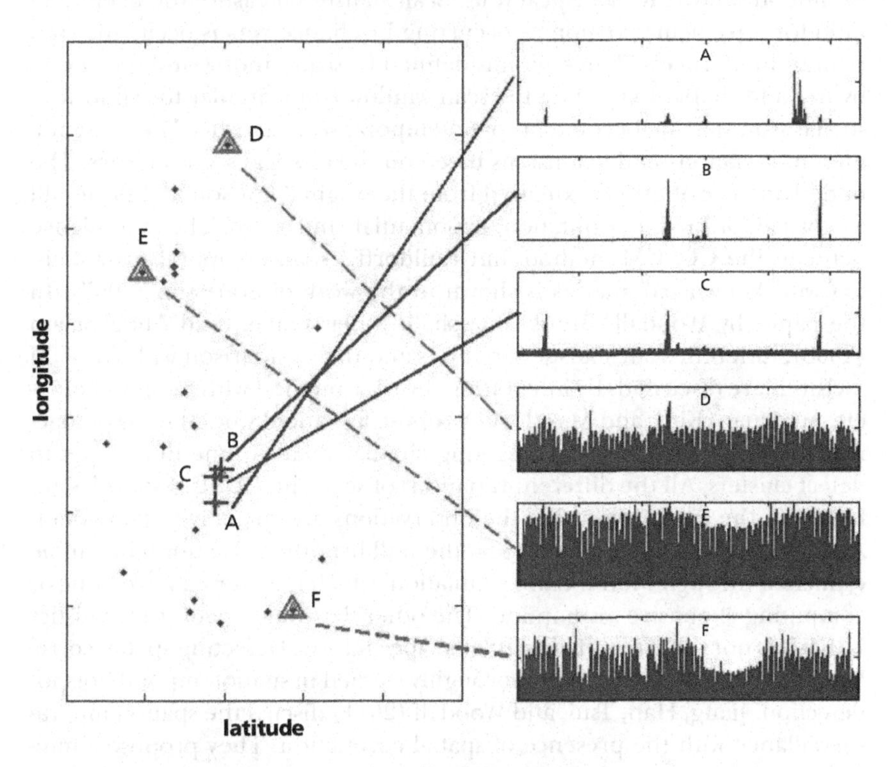


**Figure 10.3** Cell sites spatial locations and their corresponding time series. Cell sites A, B, and C are spatially y co-located; they have some common events. D, E, and F are spatially far away and there is no spatial dependency among them.

chapter is to detect the group of neighbor time series that share the similar temporal patterns. This is a spatiotemporal hotspot detection problem with spatial correlation, and the hotspot shape and locations are unknown.

Spatiotemporal hotspot detection has been well studied, and most of the basic methods are extended from original temporal or spatial approaches. Statistical process control (SPC) and time series clustering are two major approaches developed from temporal characteristics. SPC methods assume the data are temporally uncorrelated, stationary, and follow a normal distribution. Time series clustering techniques depend on the similarity definition; for example, Euclidean distance, dynamic time warping, Kullback-Leibler (KL) divergence, Pearson correlation coefficient, and so on are all used to measure similarity. Several papers (Keogh & Kasetty, 2003; Keogh & Lin, 2005; Warren Liao, 2005) provide a detailed survey regarding time series clustering. The typical and the most popular method extended from spatial approach is Kulldorff's scan statistics (Kulldorff, 1997, 2001), especially in healthcare surveillance domain. Scan statistics are essentially likelihood ratio (LR) statistical tests. Scan statistics measure the likelihood ratio for a particular region as occurring by chance versus occurring by a nonrandom process. The region is defined by scan window and it can vary by size and shape. Extending the scan window from circular to cylindrical, spatial scan statistic becomes a spatiotemporal scan statistic. There are numerous variations and extensions based on the classical scan statistics. The underlying distributions extended from the original Poisson and Bernoulli to normal, ordinal, permutation, exponential, and so on. The equivalence between the CUSUM method and Kulldorff's spatiotemporal scan statistics with known parameters is shown in the work of Sonesson (2007). In the paper by Woodall, Brooke Marshall, Joner, Fraker, and Abdel-Salam (2008), different scan statistic variations and the comparison with CUSUM methods are discussed. By integrating the scan method with temporal models, Au, Duan, Kim, and Ma (2010) propose a spatiotemporal nonhomogeneous Poisson process by aggregating circular areas to one time series to detect clusters. All the different variations of scan and CUSUM statistics are based on the assumption that the observations are spatially independent. And also, scan statistics assume that the null hypothesis is known or can be estimated through Monte Carlo simulation, which is either an unrealistic or computing-expensive assumption. The other disadvantage of scan statistics is that it is not robust to the irregular shape clusters. Detecting spatial correlated time series has not been thoroughly studied in spatiotemporal hotspot detection. Jiang, Han, Tsui, and Woodall (2011) discuss the spatiotemporal surveillance with the presence of spatial correlation. They propose a multivariate cumulative sum (MCUSUM) chart method and apply the generalized likelihood ratio test (GLRT) to develop a set of monitoring statistics given the availability of change location, time, and coverage. Following the traditional GLRT approach, the MCUSUM method assumes the observations are multivariate normal distribution, and the monitor statistics are deduced from the multivariate normal mean vector and covariance matrix.

In this chapter, we construct a multivariate dependence statistic *pseudo* $\tau$ to represent copula dependency among multiple time series. The single statistic not only relaxes the normal assumption and linear relation limitation for MCUSUM, but it also simplifies the hierarchical structure of multivariate copulas. Weighted copulas, which embed spatial dependency *pseudo* $\tau$ and temporal extreme value likelihood, are proposed to quantify the spatiotemporal irregularity. Another contribution of the chapter is a new neighborhood detection procedure, which is robust in detecting spatial irregular shapes.

The rest of the chapter is organized as follows. Section 2 provides the theoretical foundation of the work. Multivariate extreme value theory and copula concepts are briefly reviewed. *Pseudo* $\tau$ is constructed to represent various multivariate copula extensions. Weighted copulas are proposed as a measurement to detect irregular pattern temporally and spatially. Section 3 describes the procedure to detect the spatial areas that have irregular shape. Section 4 illustrates the implementation of our framework using synthetic data. Section 5 concludes the chapter with ongoing research.

## METHODOLOGY

In this section, we describe the proposed method for irregular pattern detection in spatiotemporal data. In the subsection "Multivariate Extreme Value," theory and copulas are reviewed. In the subsection "Dependence of Copulas," the relation between Kendall's and bivariate Archimedean copula are discussed. *Pseudo* $\tau$ is constructed to represent various multivariate Archimedean copula extensions. Weighted multivariate dependency that integrates the temporal characteristics and spatial dependence is proposed in that last subsection.

### Multivariate Extreme Value Theory

Extreme value theory (EVT) has been widely used in environmental applications such as rainfall, flooding, and climate modeling since the 1920s, and it has been applied to financial and insurance applications in recent years. Risk managers use EVT to estimate the risk for low probability, but catastrophic loss events. A comprehensive review of extreme value theory and applications is given in the paper by Kotz and Nadarajah (2000). Essential to EVT is the study of the stochastic tail properties of *i.i.d* random variables. For univariate extreme value distribution, there are two forms based on the way to formulate the tail. One is the block maxima, which focuses on the maximum (minimum) value distribution for a series of fixed blocks, and the other one is peak over threshold (POT), which focuses on the properties of the values over the predefined threshold. The POT approach is good at short-sequence data, but to define a proper threshold is critical. It will be either invalid the asymptotic arguments or increase the variability when the threshold is defined too low or too high.

*BH* is a block maxima measurement by definition, so we focus on the block maxima extreme value approach in this chapter. Denote $\chi^{+}_{n,m} = \max(X_{n,1},\ldots,X_{n,k},\ldots,X_{n,m})$, where $X_{n;k}$ is *i.i.d* random variables with the same distribution. If there exist sequences of constants $\{a_n > 0\}$ and $\{b_n\}$ such that $P\{(\chi^{+}_{n,m} - b_n)/a_n \le x\} \to G(x)$ as $n \to \infty$, the extreme value distribution $G(x)$ can be expressed as Equation (10.1),

$$G(x) = \begin{cases} \exp\left[-\left[1+\xi\left(\frac{x-\mu}{\delta}\right)\right]^{\frac{1}{\xi}}\right] & \text{when } \xi \neq 0 \\ \exp\left(-e^{\frac{x-\mu}{\delta}}\right) & \text{when } \xi = 0 \end{cases} \tag{10.1}$$

where

$$1+\xi\left(\frac{x-\mu}{\delta}\right) > 0, -\infty < \mu, \delta < \infty,$$

$\mu$, $\delta$, and $\xi$ are location, scale, and shape parameter respectively. When $\xi > 0$, it is Frèchet distribution, when $\xi = 0$, it is Gumbel distribution, and when $\xi < 0$, it is Weibull distribution. Extended from univariate extreme value distribution, multivariate extreme value distribution studies the dependence structure of multivariate extreme value distribution. Let $G_n$ be the marginal distribution of the univariate extreme $\chi^{+}_{n,m}$; the joint distribution $G$ of $(\chi^{+}_{1,m},\ldots,\chi^{+}_{n,m},\ldots,\chi^{+}_{N,m})$ is expressed as Equation (10.2):

$$G(\chi^{+}_{1},\ldots,\chi^{+}_{n},\ldots,\chi^{+}_{N}) = C\left(G_1(\chi^{+}_{1}),\ldots,G_n(\chi^{+}_{n}),\ldots,G_N(\chi^{+}_{N})\right) \tag{10.2}$$

where $C$ is the extreme value copula and $G_n$ is a nondegenerate univariate extreme value distribution. Correlation coefficient is the traditional way to measure the dependency. But the linearity, normality, and homoscedasticity assumptions are hard to achieve in real world applications. Copula overcomes the limitation of correlation and being well used in financial industry and climate models. *Sklar's theorem*, the foundation theorem for copulas, states that for a given joint multivariate distribution function and the relevant marginal distributions, there exists a copula function that can be expressed as:

> Let $F$ be a $d$-dimensional joint distribution with margins $F_1, F_2, \ldots, F_d$. Then there always exists a function $C: [0,1]^d -> [0,1]$ such that
>
> $$F(x_1, x_2, \ldots, x_d) = C\left(F_1(x_1), F_2(x_2), \ldots, F_d(x_d)\right) \tag{10.3}$$
>
> If $x_1, x_2, \ldots, x_d$ are continuous, then $C$ is unique; Conversely if $C$ is a copula and $F_1, F_2, \ldots, F_d$ are distribution functions, then the function $F$ is a joint distribution with margins $F_1, F_2, \ldots, F_d$.

Essentially, A $d$-dimensional copula $C$ is a multivariate distribution with uniformly distributed marginal $u(0,1)$ on $[0,1]$. Equation (10.3) can be rewritten as Equation (10.4) when $u = (u_1, u_2, \ldots, u_d)' \in [0;1]^d$

$$C(u_1, u_2, \ldots, u_d) = F\left(F_1^{-1}(u_1), F_2^{-1}(u_2), \ldots, F_d^{-1}(u_d)\right) \tag{10.4}$$

where $F_i^{-1}$ is the inverse of $F_i$.

In multivariate extreme value case, Copula $C$ is an extreme value copula. Extreme value copula properties are discussed in the paper by Gudendorf and Segers (2010). There are different types of copula to fit different problems. Gumbel copula is the only Archimedean copula that follows the extreme value distribution (Genest & Rivest, 1989), and the copula exhibits greater dependence in the positive tail than in the negative. *BH* follows extreme value distribution by definition and the positive tail, which represents the high network traffic volume as the concern in capacity planning. So Gumbel copula is the proper one in this study. Archimedean copula is a well popular copula with simple closed form expression, associative property and dependence structure. It can be expressed by generator function as in Equation (10.5).

$$C(u_1, u_2, \ldots, u_d) = \psi\left(\psi^{(-1)}(u_1) + \cdots + \psi^{(-1)}(u_d)\right) \tag{10.5}$$

The generator function for Gumbel copula is: $\psi_\alpha(x) = (-\ln(x))^\alpha$, where $\alpha$ is dependence measurement and $\alpha \in [1, \infty)$. The closed form Gumbel copula is expressed as Equation (10.6):

$$C(u_1, u_2, \ldots, u_d) = \exp\left\{-\left[(-\ln(u_1))^\alpha + \cdots + (-\ln(u_d))^\alpha\right]\right\}^{1/\alpha} \tag{10.6}$$

## Multivariate Degree of Dependence

*Kendall's* $\tau$ is a rank correlation coefficient to quantify the association between two measured quantities. The relation between *Kendall's* $\tau$ and

Archimedean copulas with generator function $\psi$ for two random variables is as in Equation (10.7):

$$\tau = 1 + 4\int_0^1 \frac{\psi(u)}{\psi'(u)} d(u) \tag{10.7}$$

*Kendall's* $\tau$ is good for bivariate, and one way to extend it to multivariate is to average the bivariate pairs. This extension basically assigns each pair the same weight and loses the differences among the pair–pair relations. The other way is to keep pair–pair dependency for multiple *Kendall's* $\tau$. But the measurement still only considers the pair–pair relations independently, and also it is hard to rank two vectors. We need to construct a statistic that is not only able to represent the relations among multiple pairs, but also easy to rank. This chapter constructs this statistic based on the properties of various techniques to extend the bivariate Archimedean copulas to multivariate Archimedean copulas.

Let's discuss the different techniques to extend the bivariate Archimedean copulas to multivariate Archimedean copulas first.

Equations (10.5) and (10.6) are natural extensions from bivariate to multivariate copula. One generator function is used to represent the multivariate dependency. This type of extension is called exchangeable multivariate Archimedean copula (EAC). There are two other types of extension: nested Archimedean construction (NAC) and pair-copula construction (PCC). Both constructions have multiple generator functions and are essential hierarchical bivariate structures based on the association property.

There are two types of NAC extensions (McNeil, 2008): fully nested Archimedean construct (FNAC) and partial nested Archimedean construct (PNAC). FNAC adds one dimension on top of previous generated copulas step by step. PNAC couples pairs of univariate distribution first and then couples pairs of the copulas of the previous step. Figure 10.4 illustrates

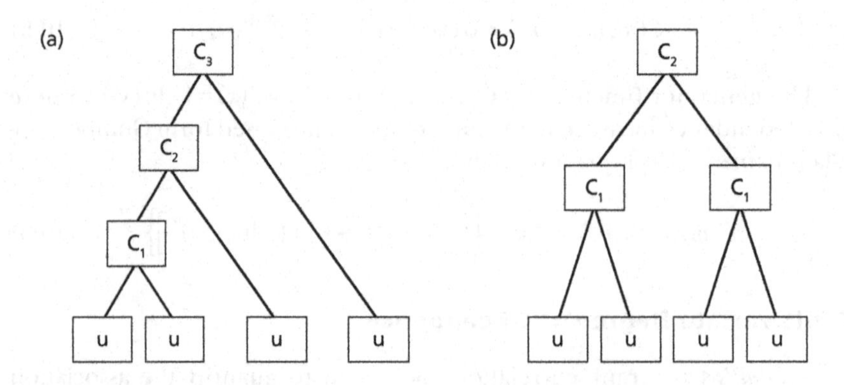


**Figure 10.4** Nested Archimedean construction: (a) Fully NAC; (b) Partial NAC.

FNAC and PNAC through a four-variable example. For $d$–dimension multivariate, NAC could model up to $d$–1 pairs of bivariate Archimedean copulas. The necessary condition for constructing NACs to be a proper multivariate copula is that the degree of dependence must decrease with the level of nesting.

PCC decomposes $d$–dimension multivariate density into $d(d-1) = 2$ bivariate copula densities, of which the first $d$–1 are unconditional, and the rest are conditional. PCC is based on the multivariate dependence concepts (Joe, 1997) and graph model (Bedford & Cooke, 2002). PCC construction, properties, and comparison are further studied in several works (Aas & Berg, 2009; Aas, Czado, Frigessi, & Bakken, 2009; McNeil & Nešlehová, 2009). PCC has D-vine and canonical-vine two forms. Figure 10.5 shows the specific decomposition methods for canonical-vine and D-Vine respectively. Canonical-vine assumes one variable to be more important than the others, and D-Vine is more symmetric. PCCs have $d(d-1) = 2$ pairs of copulas, and also the copulas are not limited to Archimedean copulas only.

Each of these extensions essentially established pairs of bivariate copulas and hieratically integrated these groups together to formulate the multivariate Archimedean copulas. The difference among the four techniques is the way to order and combine the pair copulas. The copulas in the higher level of the hierarchy depend on the cascade pairs of copulas, and the copulas on the top of the hierarchy can be considered as the prècis of all copulas. Taking advantage of this property, we define the *pseudo* $\tau^d$ to capture the multivariate hierarchical dependency relation.

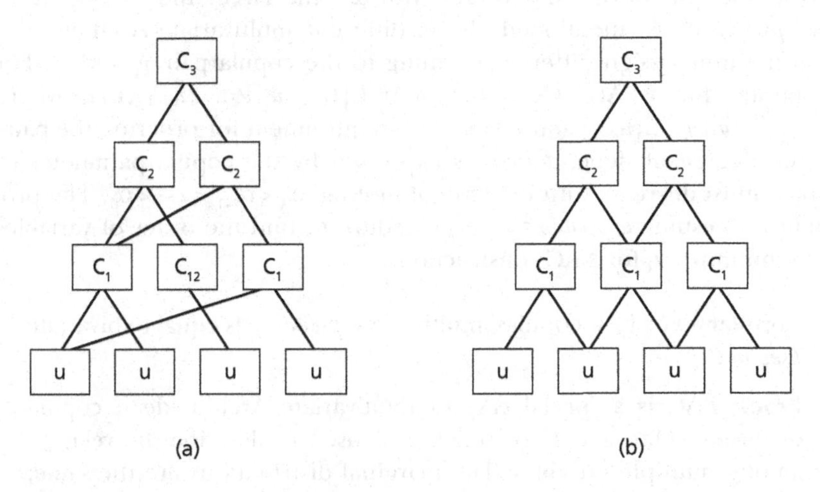


**Figure 10.5** Pair copula construction: (a) PCC—Canonical Vine; (b) PCC—D-Vine.

**Lemma**. For a $d$–dimension and $k$–level Archimedean copula, the relation of copula $C^d(u_1, u_2, \ldots, u_d)$ and *pseudo* $\tau^d$ can be simplified as

$$\tau^d = 1 + 4\int_0^1 \frac{\psi^k(t)}{\psi^{k'}(t)} d(t) \tag{10.8}$$

where $\psi^k(t)$ is the $k$th level generator function.

**Proof**. The proof is straightforward from Copula's definition, its association property, and the method to construct multivariate copula. The copula itself is a multivariate distribution with uniformly distributed marginal $u(0, 1)$ on $[0,1]$, for each step of multivariate construction generates a new multivariate distribution that is uniformly marginal, and this new multivariate distribution can be treated as a new element to construct the next level copula. Let $v_1 = C_1^{k-1}$ and $v_2 = C_2^{k-1}$

$$C^d(u_1, u_2, \ldots, u_d) = C(C_1^{k-1}, C_2^{k-1}) = C(v_1, v_2) \tag{10.9}$$

The *pseudo* $\tau^d$ for $d$–dimension multivariate is essential a bivariate $\tau$ between random variables $v_1$ and $v_2$, Since the generator of $C(v_1, v_2)$ is $\psi^k$, Hence *pseudo* $\tau^d$ follows Equation (10.8). For Gumbel family, *pseudo* $\tau^d$ can be expressed as in Equation (10.10):

$$\tau^d = 1 + 4\int_0^1 \frac{tint}{\alpha^k} d(t) = 1 + \frac{4}{\alpha^k}\left(\left[\frac{t^2}{2} lnt\right]_0^1 - \int_0^1 \frac{t}{2} dt\right) = 1 - \frac{1}{\alpha^k} \tag{10.10}$$

Gumbel *pseudo* $\tau^d$ is monotonic with $\alpha^k$—the larger the $\alpha^k$, the larger the *pseudo* $\tau^d$. As mentioned above, different multivariate Archimedean constructions assign different meaning to the copula pair $v_1$ and $v_2$. For example, for FNAC, $C^d(u_1, u_2, \ldots, u_d) = C\left(C^{d-1}(u_1, u_2, \ldots, u_{d-1}), u_d\right)$, where $v_1 = C^{d-1}(u_1, u_2, \ldots, u_{d-1})$ and $v_2 = u_d$. The requirement for ordering the pairs is the degree of dependence, as expressed by the copula parameter $\alpha$, which must decrease with the level of nesting: $\alpha_k < \alpha_{k-1} < \cdots < \alpha_1$. The procedure to estimate *pseudo* $\tau^d$ is a procedure to find the order of variables that minimize $\alpha_k$ for NAC construction.

**Corollary**: For EAC copulas, multivariate *pseudo* $\tau^d$ is equal to bivariate *pseudo* $\tau$.

**Proof**: EAC is a special case of multivariate Archimedean copula extension. One generator function is used to describe the relation among multiple variables. The marginal distributions are the same, $\theta_d = \theta_{d-1} = \cdots = \theta_1$ and $\tau^d = \tau$, regardless of the dimension $d$.

For example, using FNAC to expend a 3-dimension Gumbel family copula is expressed as in Equation (10.11):

$$C^3(u_1,u_2,u_3)=\exp\left\{-\left(\left[(-lnu_1)^{\alpha 1}+(-lnu_2)^{\alpha 1}\right]^{\frac{\alpha 1}{\alpha 2}}(-lnu_3)^{\alpha 2}\right)^{\frac{1}{\alpha 2}}\right\} \quad (10.11)$$

and $\tau^3 = 1 - 1/\alpha_2$, if $\alpha_1 = \alpha_2$, Equation (10.11) becomes EAC and $\tau^3 = 1 - 1/\alpha_2 = 1 - 1/\alpha_1$.

The advantage of *pseudo* $\tau^d$ defined in Equation (10.8) is that the single number multivariate dependence measurement can compare the dependency among copulas with different dimensions. Also, Equation (10.8) represents the dependency among all time series, not only pair–pair relations. For EAC, dependency is determined by only one generation function and there is only one parameter. For hierarchical construction, dependency is always measured by the last pair of copulas regardless of the dimension. The simplicity of Equation (10.8) provides a straightforward measurement to evaluate the dependency among multiple time series, which can be used to detect the spatial correlated time series in hotspot detection.

After constructing the *pseudo* $\tau^d$ to measure the multivariate dependence, let's discuss how to utilize this measurement to identify the temporal groups that have different behaviors compared with its neighbors.

## Weighted Spatial Dependency Model

Here we construct *pseudo* $\tau^d$ to measure the dependency for multivariate time series. Spatial dependency can be tested by the change of *pseudo* $\tau^d$ when involving different numbers of time series. The detailed procedure will be illustrated in the next section, but *pseudo* $\tau^d$ only measures the spatial dependency among the time series without considering the degree of temporal skewness. To detect groups that have large different temporal traffic patterns compared with their neighbors, weighted dependency is proposed. Weighted dependency integrates temporal likelihood and spatial dependency to represent the area with irregular patterns.

As discussed in the first section, the temporal feature for each univariate time series is an extreme value distribution, and its *log-likelihood* function is:

$$\log \boldsymbol{L}(\theta;\boldsymbol{X})=\sum_{i=1}^{n}\log G(x_i,\theta) \quad (10.12)$$

A regular steady traffic time series is constructed by aggregating all time series in a large region *A*. The aggregated time series smooth all the local events and only show the common patterns due to the aggregation mask effect. The large region can be at zip code level, county level, or state level. The aggregated time series also follows the extreme value distribution with parameter $\theta^A$.

The log form likelihood ratio between any site $i$ and the aggregated series is expressed in Equation (10.13):

$$ll_i = \log \boldsymbol{L}(\theta_i, \boldsymbol{X}) - \log \boldsymbol{L}(\theta^A, \boldsymbol{X}) \tag{10.13}$$

When the network traffic for site $i$ is similar as the normal traffic, $ll_i$ is small, Otherwise, $ll_i$ will be big. Assume an area $A$ covers $p$ time series with GEV parameters $[\theta_1, \cdots, \theta_p]$, The weighted dependence for the area $A$ is defined as Equation (10.14):

$$W\tau^A = \frac{1}{p}\sum_{i=1}^{p}(ll_i) * \tau^A \tag{10.14}$$

For Gumbel family, Equation (10.14) can be expressed with copulas parameter $\alpha$, as in Equation (10.15):

$$WC^A = \frac{1}{p}\sum_{i=1}^{p}(ll_i) * \left(1 - \frac{1}{\alpha^A}\right) = \frac{1}{p}\sum_{i=1}^{p}\left(\log L(\theta_i, X) \log L(\theta^A; A)\right) * \left(1 - \frac{1}{\alpha^A}\right) \tag{10.15}$$

The spatial correlated spatiotemporal irregular pattern detection problem is transferred to a problem to find the area $A$ that has the largest $WC^A$.

## ALGORITHM

In this section, we demonstrate the procedure to identify the irregular shaped hotspot based on multivariate weighted dependency. There are two major steps: temporal feature extraction and spatial neighborhood detection. Temporal feature extraction estimates the time series features as univariate time series. Spatial neighborhood detection evaluates the neighborhood dependency and finds the group that has local maximum-weighted copulas. The spatial detection method is robust to handle the situation when the neighbor has an irregular spatial shape. The spatial method involves five major steps: initialization, estimation, adaptation, expansion, and finalization. The details of each step are described as following.

### Temporal Feature Extraction

As introduced in Section 1, generalized extreme value distribution is a proper model to extract tail properties that emphasize the low frequency but high traffic load events. GEV is a suitable model to extract BusyHour time series characteristics. Due to the multiple periodic structures of the original hourly time series, preprocessing is needed to remove the periodic patterns and normalize the time series before constructing BusyHour time series. For an $N \times p$ dataset, $p$ is the number of cell site and $N$ is the length of BusyHour. Each time series is modeled as GEV with parameters $\theta_i = [\mu_{i,} \delta, \xi]$. $i \in [1{:}p]$, and the likelihood ratio with the standardized aggregate time series is

$$ll_i = \frac{\theta_i}{\theta_0},$$

as specified above. The temporal feature extraction step can be summarized as:

1. Remove periodic pattern and normalize the original time series to construct BusyHour time series.
2. Estimate parameters $\theta_i$ for each BusyHour time series by GEV model.
3. Calculate the likelihood ratio $ll_i$ for each BusyHour time series versus the aggregated time series.
4. Rank all the BusyHour time series based on likelihood ratio $ll_i$.

## Spatial Neighborhood Detection

After extracting the temporal features, the next step is to evaluate the neighborhood relation. Since we are more interested in the site that is different from the normal sites, the second step starts from the time series that has the largest likelihood ratio. Consider this particulate time series as the first seed.

The five-step procedure that detects the neighborhood relation based on weighted multivariate dependence is illustrated as following:

1. Initialization: Let $i$ denote the location of the selected seed cell site and $A_i^r$ is the area that covers a region with center $i$ and radius $r$. The number of cell sites in $A_i^r$ is $p_i^r$. Initial $r$ as small as possible, but $p_i^r \geq 2$. The time series in $A_i^r$ are considered as multivariate time series.
2. Estimation: The multivariate Gumbel copula and weighted Gumbel dependence are calculated based on Equations (10.10) and (10.15) for region $A_i^r$, denoted as $C_i^r$, and $WC_i^r$, respectively. Iteratively compute $WC_i^r$ with $r = r + \Delta r$ as the iteration number increases until there is no increment of weighted copula. $\Delta r$ is the size of predefined incremental radius.

3. Adaptation: This step adjusts the border sites and evaluates the site that has strong similarity with the identified group or not. Assume the $r$ stops at the above step, $r^s$, the cell sites that included in $A_i^{r^s}$, but not in $A_i^{(r^s-\Delta r)}$ denotes as $A_i^{r^s}$. Compute the weighted dependence W $WC_{ij}^{rs}$ for the area $\{A_i^{(r^s-\Delta r)} \cup j\}$, where $j$ is individual cell site in $A_i^{r^s}$. For all $j$ with $WC_{ij}^{r^s} \geq WC_i^{r^s}$ will be included in the dependency neighbor centered with $i$ noted as $\widehat{A_i}$, and its weighted copula is $WC^{\widehat{A_i}}$, $\widehat{A_i}$, is not necessarily a circle now.
4. Expansion: For each site that is covered in the region $\widehat{A_i}$ exclude $i$, go back to the stage estimation and adaptation to achieve a series of $\widehat{A_{ij}}$ and the weighted copula is $WC^{\widehat{A_{ij}}}$.
5. Finalization: if $WC^{\widehat{A_{ij}}} > WC^{\widehat{A_i}}$, the new neighbor is redefined as $\mathrm{A_i} = \{\widehat{A_{ij}} \cup \widehat{A_i}\}$. The group series that have high spatial dependence and large difference from the normal traffic are detected.

The detected neighborhood cluster is not necessary a circle. The robust of irregular spatial shape is more accurate than circle detection. All the covered sites are labeled, the process select the next non-labeled largest likelihood ratio sites as the next seed and repeat the initialization–estimation–adaptation–extension–finalization to achieve the neighborhood region for each seed. The whole process continues until the difference of weighted copulas is less than a threshold $T$.

## EXPERIMENTS ON SYNTHETIC DATA

To validate the proposed method, we generate $N \times p$ data matrix to simulate the real network traffic pattern $p = 100$ and $N = 3360$ (140 days × 24 hours on each day). We adopt the nonhomogeneous Poisson process (NHPP) to simulate the normal traffic. NHPP is a well-developed and widely used temporal model to capture the network traffic characteristics (Fischer & Meier-Hellstern, 1993; Ihler, Hutchins, & Smyth, 2006; Kuhl, Wilson, & Johnson, 1997; Massey, Parker, & Whitt, 1996). The observed time series are generated by adding event effects on top of the normal traffic, and the detail follows Au et al.'s (2010) paper. One thing that needs to be emphasized is that the simulation model focused on generating the network traffic data should be able to represent the real data without considering the multivariate extreme value and Copulas that have been used in the solution model. The synthetic data are generated by totally different mechanisms from the proposed detecting model.

Three irregular regions are designed with different patterns that have 10, 20, and 50 events, respectively. Figure 10.6(a) shows the spatial relation of the experiment data. The x-axis and y-axis indicate the latitude and

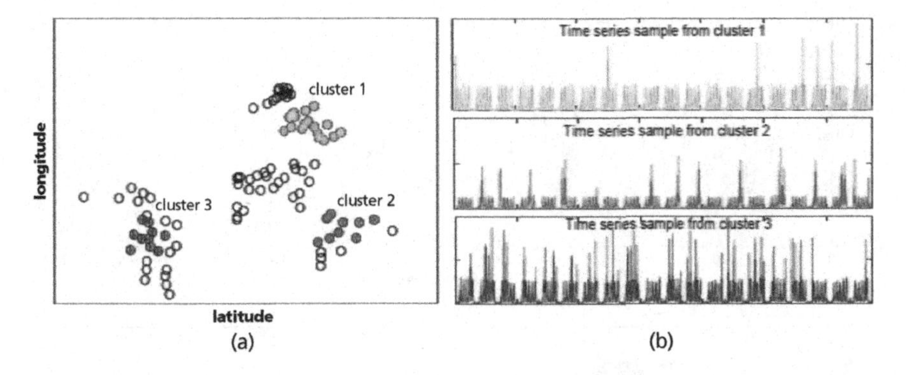


**Figure 10.6** (a) Spatial display of the three venue site clusters. Three venue sites are indicated by three different colors: red, blue and green. (b) Sample traffic volume time series from each cluster.

longitude. The empty circles represent the locations of cell sites, and the three groups of star circles are three clusters that have 10, 20, and 50 events, respectively. The time series data for the three clusters are shown in Figure 10.6(b), two samples from each group.

Extract the BusyHour series by removing periodic and standardization as described above; the BusyHour series and the histogram are shown in Figure 10.7, one example from each cluster. Figure 10.7(a) is the time series after removing the periodic and standardization. Figure 10.7(b) is the histogram. Figure 10.8 is the aggregated BusyHour series and histogram for all 100 simulated data. Comparing three clusters with the aggregate data,

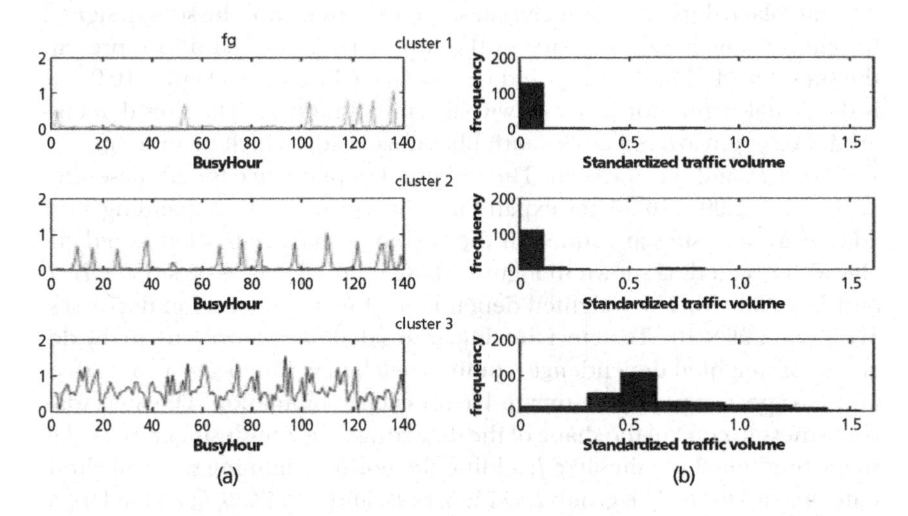


**Figure 10.7** Standardized BusyHour and histogram for three clusters.

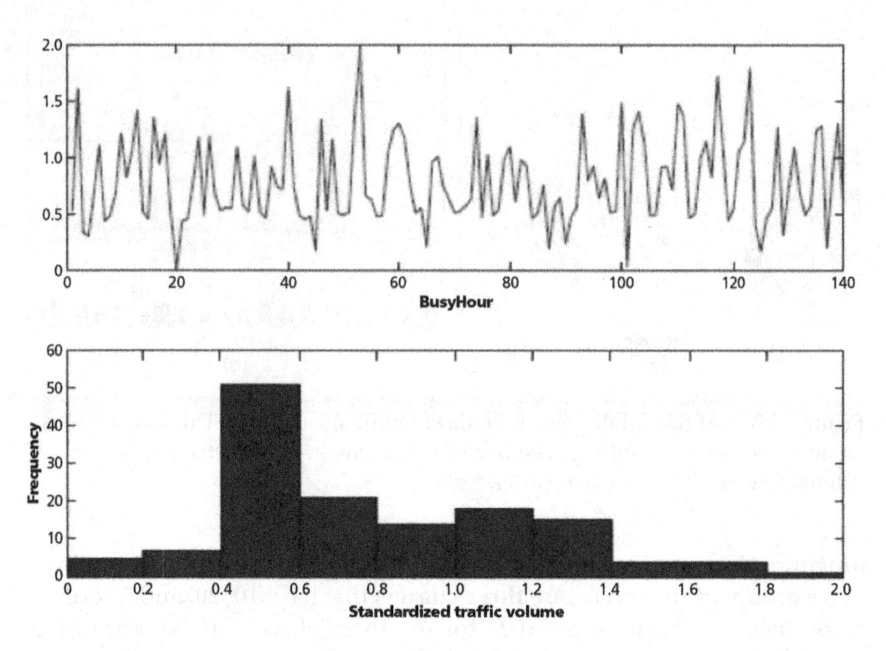


**Figure 10.8** Aggregated BusyHour(BH) and histogram.

all of the three clusters have heavier tails than the aggregated data, and Clusters 1 and 2 have heavier tails than cluster 3. This is in sync with the design, where the three clusters have 10, 20, and 50 events separately. Apply the proposed method described in Table 10.1 to the simulated data set.

Figure 10.9 illustrates the procedure to detect Cluster 2. In each subfigure, the black dots represent cell sites; the red circles are the sites designed for cluster, which has 20 events in the time series. The red dots represent the sites detected by the proposed method for Cluster 2. In Figure 10.9, (a) is the initial region for seed "A" with the initial radius $r$. The sites that covered the region are red circles with blue dots inside, which include sites: $A$, $B$, $C$, and $D$, and $A$ is the seed. The weighted dependence for all these sites is $WC_{step(a)} = 2.99 \times 10^7$. After expanding one more step by expanding $r$ for adding $\Delta r$, new sites are added in the region, which are labeled as red circles with green dots shown in Figure 10.9(b), and five new sites, $E$, $G$, $H$, $I$, and $J$, are added. The weighted dependence for the new region decreases, $WC_{step(b)} = 2.96 \times 10^7$. To detect the time series that one contributes to the decrease of weighted dependence, combine each new added site with $A$, $B$, $C$, and $D$ respectively and re-estimate $WC$ for each combination. This step adds robustness to control the shape of the detecting area. Only using the circular shape might include sites like $J$; adding the border trimming step will eliminate this situation. The group $E \cup \{A, B, C, D\}$ and $G \cup \{A, B, C, D\}$ has larger $WC$ than $WC_{step(a)}$; the updated detected sites are $\{E, G\} \cup \{A, B, C, D\}$, which

**TABLE 10.1 Procedure for Detecting Irregular Site**

**Initial**

1. set detected cite *label* = 0
2. set weighted copula increment threshold $d^{wc} = T$
3. set initial radius $r$ and radius increment $\Delta r$

**Normal Site Feature Estimation**

4. $S^R = \Sigma$(time series in the region $R$)
5. fit GEV model, obtain GEV parameters $\theta_0 = \text{GEV}(S^R)$

**Temporal Feature Extraction**

6. **for** (i in 1 to p, the number of all cell sites (time series))
7. estimate GEV parameters and likelihood $[\theta_i, ll_i] = \text{GEV}(S^i)$
8. calculate likelihood $ll_0 = \text{GEV}(S^i, \theta_0)$
9. calculate likelihood ratio $ll_{ri} = ll_i / ll_0$
10. **end for**
11. calculate $WC$ for all time series: $WC = \frac{1}{p}\sum_{i=1}^{p}(ll_{r_i}) * C(S)$
12. pick the seed time series: seed = $\max(ll_{ri})$, where $label_i = 0$
13. **while** ($d^{wc} >= T$)
14. initial region $A = \Phi$
15. find trimmed region around $i$: $A_i = FindRegion(i, A)$

**Extension**

16. **for** (m in $A_i$)
17. find trimmed region around m: $A_m = FindRegion(m, A_i)$
18. **end for**

**Finalize the Region**

19. final region for $i$: $A_i = A_i \cup A_m$, where $WC^{A_m} > WC^{A_i^r}$
20. Calculate the final region weighted copula: $WC^{A_i}$
21. $d^{wc} = WC - WC^{A_i}$
22. **end while**

**Subroutine: FindRegion (i, A)**

**Calculate neighborhood weighted copula of the seed time series**

1. $d^{WC_i} = 0$
2. $A_i^r = A_i^r \cup A$
3. **while** ($d^{WC_i} > 0$)
4. estimate $WC^{A_i^r}$
5. $r = r + \Delta r$
6. estimate $WC^{A_i^{(r+\Delta r)}}$
7. $d^{WC_i} = WC^{A_i^r} - WC^{A_i^{(r+\Delta r)}}$
8. **end while**

**Trim the border sites:**

9. $ds_i = diffset(A_i^{r+\Delta r}, A_i^r)$
10. **for** ($j$ in $ds_i$)
11. $A_{ij}^r$ = add $j$ to area $A_i^r$
12. estimate $WC^{A_{ij}^r}$
13. **end for**
14. Adjusted region $A_i = A_i^r \cup j$, where $WC^{A_{ij}^r} > WC^{A_i^r}$

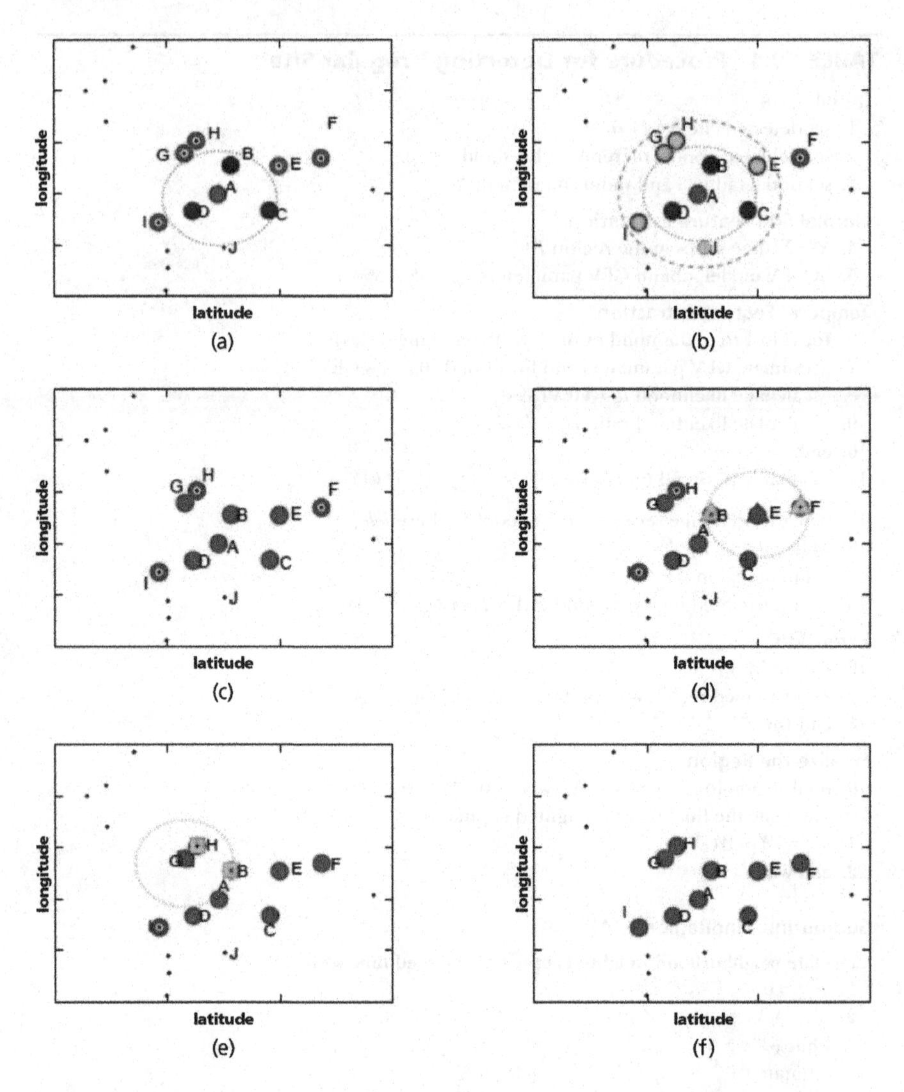


**Figure 10.9** Iteratively detect procedure. Small black dots represent the cell sites, and the red empty circles are the cell sites designed as irregular sites. Red star circles are the cell sites that have been detected (a) Initial region for seed "A" and initial radius $r$; (b) Extended region $r + \Delta r$ by seed "A"; (c) Final region by seed "A"; (d) Extended region by seed "E"; (e) Extended region by seed "G"; (f) Final region.

are the red dots in Figure 10.9(c). The weighted dependence for the newly detected region is $WC_{step(c)} = 3.32 \times 10^7$. Consider each detected site in (c) as new seeds respectively and repeat steps (a) and (b). Figure 10.9(d) shows the result for setting $E$ as the new seed, and its circle region includes sites

**TABLE 10.2 Step-Wise Result for Detecting Cluster 2**

| | Covered Cell Sites | Weighted Copula |
|---|---|---|
| Step (a) | {A, B, C, D} | 2.99 × 107 |
| Step (b) | {A, B, C, D} ∪ {E, G, H, I, J} | 2.96 × 107 |
| Step (c) | {A, B, C, D, E, G} | 3.32 × 107 |
| Step (d) | F ∪ {A, B, C, D, E, G} | 3.35 × 107 |
| Step (e) | H ∪ {A, B, C, D, E, G, F } | 3.38 × 107 |
| Step (f) | {A, B, C, D, E, G, F, H, I} | 3.41 × 107 |

labeled as red and green triangle $\{E, B, F\}$. Figure 10.9(e) is similar to (d). Its seed is $G$ and its new added site is $H$. Repeat the steps for each detected site, and the final region for Cluster 2 is the red dots covered area shown in Figure 10.9(f). The step by step result is illustrated in Table 10.2.

Figure 10.10 shows the final result for detecting the whole dataset. Comparing it with the original design as in Figure 10.6(a), the proposed method successfully detects Clusters 1 and 2 without mistake. For Cluster 3, there are two sites marked as blue cross circles that are misclassified in Cluster 3. This is in sync with the design that the time series in Cluster 1 and 2 have longer tails than Cluster 3. Cluster 3 is more like normal traffic than the other two clusters, which increases the challenge to identify Cluster 3.

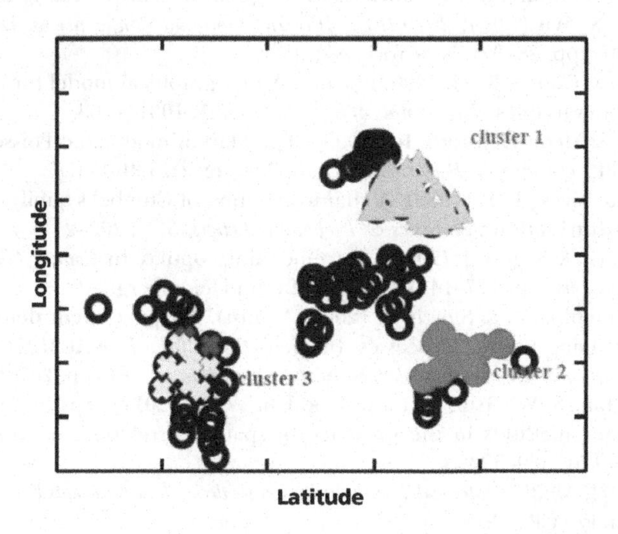


**Figure 10.10** Detected event clusters. Three event clusters are indicated by three different symbols: star, cross, and triangle. Seven filled circles indicate the cells with two events. For each cluster, the center point is marked by "X."

## CONCLUSION

This chapter proposes the multivariate extreme value concept in detecting irregular patterns from spatial and temporal datasets. Multivariate dependence statistic *pseudo* $\tau^d$ is constructed to measure the spatial relationship for $d$ time series. Weighted dependency is introduced to integrate the spatial dependency and temporal extreme value characteristics. A new spatial scanning procedure is presented to detect the hotspot with an irregular shape. The proposed method does not only consider the spatial dependency in spatiotemporal irregular pattern detection and relax the normality and linearity assumptions, but it also illustrates a new spatial scanning procedure to detect irregular shapes robustly. The proposed method performs very well on synthetic data.

## REFERENCES

Aas, K., & Berg, D. (2009). Models for construction of multivariate dependence–a comparison study. *The European Journal of Finance, 15*(7–8), 639–659.

Aas, K., Czado, C., Frigessi, A., & Bakken, H. (2009). Pair-copula constructions of multiple dependence. *Insurance: Mathematics and Economics, 44*(2), 182–198

ATIS. (2015). American National Standard Telecom. Retrieved from http://www.atis.org/glossary/

Au, T. S., Duan, R., Kim, H., & Ma, G. Q. (2010, December). Spatiotemporal event detection in mobility network. In G. I. Webb, B. Liu, C. Zhang, D. Gunopulos, & X. Wu (Eds.), *2010 IEEE 10th International Conference on Data Mining (ICDM)* (pp. 28–37). New York, NY: IEEE.

Bedford, T., & Cooke, R. M. (2002). Vines: A new graphical model for dependent random variables. *The Annals of Statistics, 30*(4), 1031–1068.

Fischer, W., & Meier-Hellstern, K. (1993). The Markov-modulated Poisson process (MMPP) cookbook. *Performance Evaluation, 18*(2), 149–171.

Genest, C., & Rivest, L. P. (1989). A characterization of Gumbel's family of extreme value distributions. *Statistics & Probability Letters, 8*(3), 207–211

Gudendorf, G., & Segers, J. (2010). Extreme-value copulas. In *Copula Theory and Its Applications* (pp. 127–145). Springer Berlin Heidelberg.

Ihler, A., Hutchins, J., & Smyth, P. (2006, August). Adaptive event detection with time-varying Poisson processes. In *Proceedings of the 12th ACM SIGKDD international conference on Knowledge discovery and data mining* (pp. 207-216). ACM.

Jiang, W., Han, S. W., Tsui, K. L., & Woodall, W. H. (2011). Spatiotemporal surveillance methods in the presence of spatial correlation. *Statistics in medicine, 30*(5), 569–583.

Joe, H. (1997). *Multivariate models and multivariate dependence concepts* (Vol. 73). Boca Raton, FL: CRC Press.

Keogh, E., & Kasetty, S. (2003). On the need for time series data mining benchmarks: A survey and empirical demonstration. *Data Mining and knowledge discovery,* 7(4), 349–371.

Keogh, E., & Lin, J. (2005). Clustering of time-series subsequences is meaningless: implications for previous and future research. *Knowledge and information systems, 8*(2), 154–177.

Kotz, S., & Nadarajah, S. (2000). *Extreme value distributions: Theory and applications.* London: Imperial College Press.

Kuhl, M. E., Wilson, J. R., & Johnson, M. A. (1997). Estimating and simulating Poisson processes having trends or multiple periodicities. *IIE transactions, 29*(3), 201–211.

Kulldorff, M. (1997). A spatial scan statistic. *Communications in Statistics-Theory and Methods, 26*(6), 1481–1496.

Kulldorff, M. (2001). Prospective time periodic geographical disease surveillance using a scan statistic. *Journal of the Royal Statistical Society: Series A (Statistics in Society), 164*(1), 61–72.

Massey, W. A., Parker, G. A., & Whitt, W. (1996). Estimating the parameters of a nonhomogeneous Poisson process with linear rate. *Telecommunication Systems, 5*(2), 361–388.

McNeil, A. J. (2008). Sampling nested Archimedean copulas. *Journal of Statistical Computation and Simulation, 78*(6), 567–581.

McNeil, A. J., & Nešlehová, J. (2009). Multivariate Archimedean copulas, d-monotone functions and $\ell_1$-norm symmetric distributions. *The Annals of Statistics,* 3059–3097.

Sonesson, C. (2007). A CUSUM framework for detection of space–time disease clusters using scan statistics. *Statistics in Medicine, 26*(26), 4770–4789.

Warren Liao, T. (2005). Clustering of time series data—a survey. *Pattern Recognition, 38*(11), 1857–1874.

Woodall, W. H., Brooke Marshall, J., Joner Jr., M. D., Fraker, S. E., & Abdel-Salam, A. S. G. (2008). On the use and evaluation of prospective scan methods for health-related surveillance. *Journal of the Royal Statistical Society: Series A (Statistics in Society), 171*(1), 223–237.

CHAPTER 11

# ROAD SAFETY DETECTION MODELING BASED ON VEHICLE MONITORING DATA IN CHINA

**Xing Wang, Wei Yuan**
***Renmin University of China***

**Susan X. Li**
***Adelphi University***

**Zhimin Huang**
***Adelphi University and***
***Beijing Institute of Technology***

## ABSTRACT

By investigating information of an average speed of vehicles and characteristics of speed distribution and speed volatility, this chapter develops a road safety detection system model in relating to real-time speed. This model analyzes the risk degree across double dimensions of roads and vehicles. The dimension of roads focuses on characters of speed dispersion and distribu-

*Contemporary Perspectives in Data Mining, Volume 2*, pages 215–228

tion, constructs coefficients of variation for different roads, and recognizes multimodal distributions through nuclear density. The dimension of vehicles focuses on characteristics of speed changing in moving succession and analyzes outliers of speed, speed clusters, and speed variance in adjacent roads. The calculating outcomes of both above dimensions are combined to be "risk factors" for road safety, which can judge risk degree of both road and vehicle, as well as issue real-time warning for road and vehicle. The empirical results show that this model provides a strong theoretical support for road safety detection based on vehicle monitoring data.

## INTRODUCTION

According to statistical data, a total of 210,812 road traffic accidents occurred in China in 2011; 237,421 people were injured, and 62,378 were killed. The number of deaths due to traffic accidents in China has ranked first in the world for ten consecutive years. Therefore, it is particularly important that how to make real-time and accurate road safety detection to reduce the incidence of traffic accidents.

The theory of detection has applied seldom in road safety in the literature. The current index system in China mainly focused on statically historical data. For example, Shao (2005) constructed an index system on safety detection of urban traffic in 2005 using individual attributes of the driver, characteristics of vehicle performance, road safety, and traffic management; Feng and Liu (2008) constructed a detection system of traffic accidents by means of sensor technology, which was established by the collection, transmission, processing, and output of information. From macroscopic and microscopic aspects, Wang and Liu (2010) established an index system of road safety detection based on the causes of urban traffic accidents in different road sections. The major consideration of the above systems is statically historical data that include the educational background of the driver, the vehicle performance, and the incidence of traffic accidents. Those stative data are limited when to recognize road risk in vehicle traveling process. Furthermore it is difficult to measure risk factors. The average speed is often used as a single index in the research of risk detection. Speed has a great influence on road safety, and the dispersion and volatility of speed distribution are important indicators for measuring road safety (Ma, Liu, & Zheng, 2009). Thus, based on the two dimensions of road and vehicle, the chapter develops a real-time system model of road safety detection through analyzing the distributions, characteristics, and volatility of speed data and provides a more detailed method to seek risk factors and a new research direction of road safety detection.

## THE FRAMEWORK OF ROAD SAFETY DETECTION SYSTEM

Figure 11.1 shows a road safety detection system based on real-time speed. The system consists of three parts. The first part is data collection, collecting the real-time speed data of moving vehicles via the sensor technology. The second part is data processing, which analyzes the risk degree of real-time speed data from the two dimensions of road and vehicle and establishes indicators of risk for each road section and constructs risk factors. The dimension of road focuses on characteristics of speed dispersion and distribution and constructs coefficients of variation for different road sections and recognizes multimodal distributions through nuclear density. The dimension of vehicle focuses on characteristics of speed variance in driving succession and investigates speed variance in adjacent road sections after classifying speed data and processing outliers of speed. The outcomes of both above dimensions are combined to be "risk factors" for road calculation. The third part is the application of outputs. We will conduct the risk judgment for road and vehicles by means of risk factors and issue real-time warnings for road sections and vehicles that are considered dangerous. The plan should have a good real-time performance and a strong predictive ability to make full use of real-time speed data based on dynamic indicators and comprehensive and reliable outputs.

## THE CONSTRUCTION OF SPECIFIC INDICATORS

### Construction of Detection Indicators from the Dimension of Roads

According to research by Wu and Wu (2008), there is a positive correlation between the variable coefficient of speed and the incidence of traffic accidents—namely, that the larger the variable coefficient of speed is, the greater the incidence of traffic accidents is. Thus, we introduce coefficient of variation (standard deviation/mean value) as one of the indicators for road safety detection. The larger the coefficient of variation is, the more possible traffic accidents occur and the higher the detection level is. Otherwise, it is relatively safe.

Through the average speed data of the current 19 vehicles on 20 road sections between Beijing and Beidaihe, we calculate the coefficient of variation for each road section and array the results in ascending order (see Table 11.1). It shows that the variable coefficients of each road section differ. Among them, road section s4 has the smallest coefficient of variation, 0.06, which shows that this section has a better real-time safety, while section s16 and s12 have the largest coefficient of variation, 0.380733, which shows

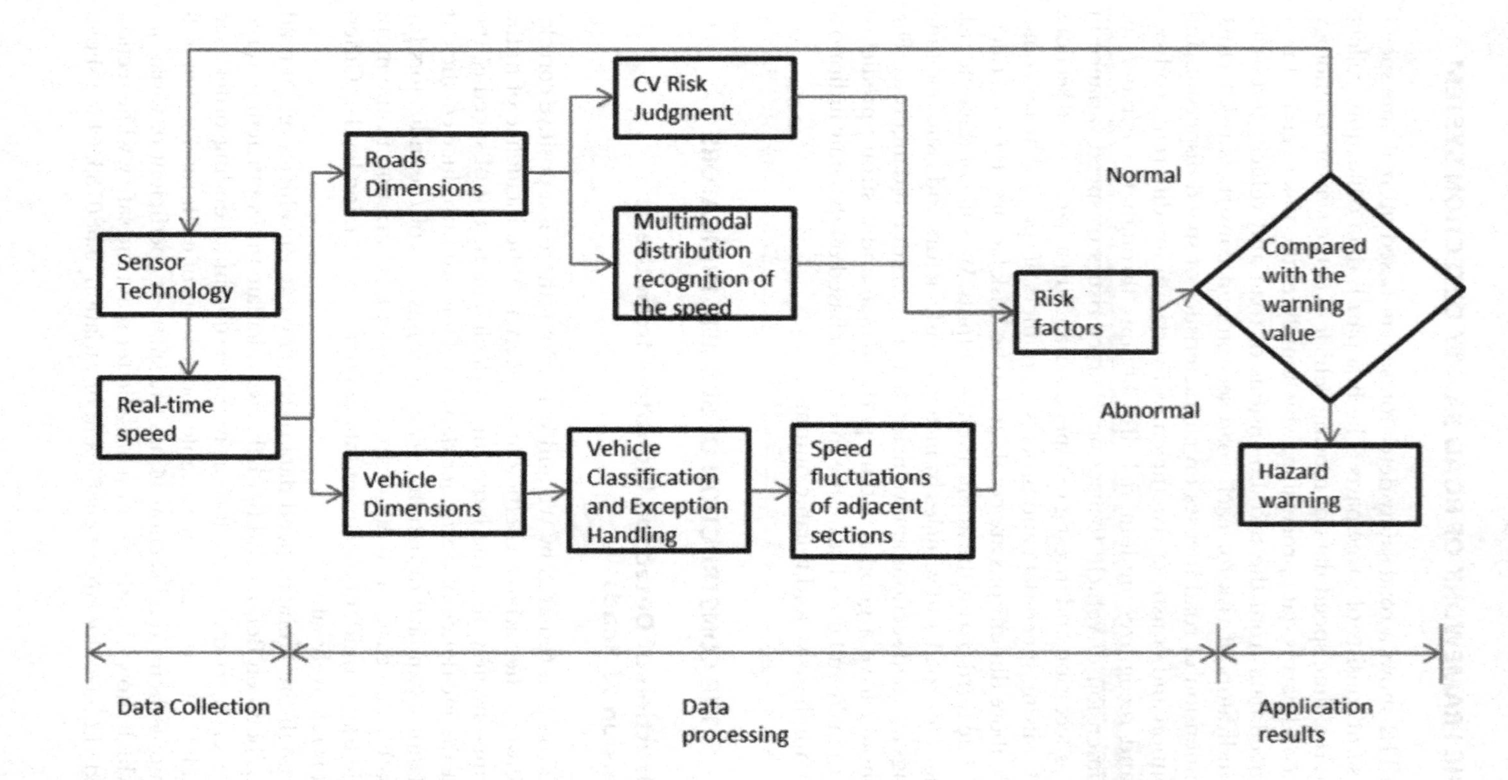


**Figure 11.1** The framework of road-safety detection system.

**TABLE 11.1 Variable Coefficients of Each Road Section**

| Sections | s4 | s17 | s11 | s14 | s7 | s13 | s10 | s19 | s6 | s3 |
|---|---|---|---|---|---|---|---|---|---|---|
| Coefficient Variance | 0.06 | 0.08 | 0.10 | 0.12 | 0.13 | 0.17 | 0.21 | 0.23 | 0.25 | 0.25 |
| Sections | s8 | s15 | s20 | s9 | s5 | s1 | s18 | s2 | s16 | s12 |
| Coefficient Variance | 0.25 | 0.25 | 0.27 | 0.27 | 0.30 | 0.30 | 0.33 | 0.36 | 0.38 | 0.38 |

that the two sections have the worst real-time safety and a higher possibility of traffic accidents.

Based on the existing data, we arrange the standard deviation of speed on each road section in correspondence to the ascending order of average speed, aiming to further explore the law between speed and the variance (see Figure 11.2). From Figure11.2, we can see that as the average speed increases, the standard deviations of different speeds present a trend of increase at first and then decrease later. To verify this claim, we divide the data into two parts at the data peak and use the method of Cox–Staut to conduct the tendency test for the two parts respectively: whether the data of first half has an increase trend and the second half has a decrease trend. The test results of P are 0.5 and 0.03125, respectively. When $\alpha = 0.05$, the first half has no increase trend, while the second half has a distinct decrease trend.

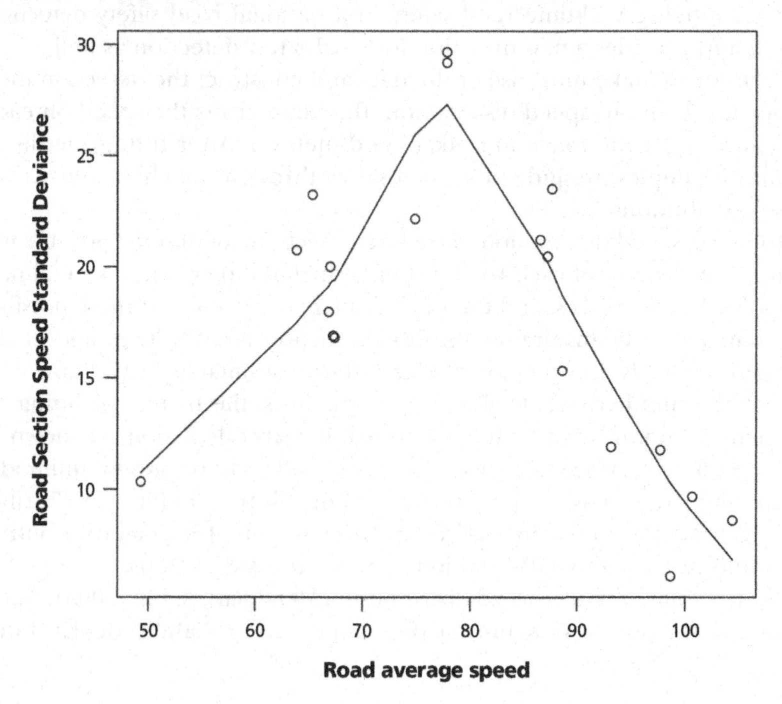


**Figure 11.2** Variance and speed relationship.

It shows that, when the average speed is less than the critical value, as the average speed increases, the standard deviation of the average speed has no distinct increase trend (the tendency existing in the test of Cox–Staut failed in our experiment, but the small number of samples is a possible reason for the failure); when the average speed reaches the critical value (about 80 km/h), the standard deviation of the average speed reaches its maximum; when the average speed is greater than the critical value, as the average speed increases, the standard deviation of the average speed has a distinct decrease trend. In the area marked by grey box on Figure 11.2, the mean value is medium, while the standard deviation is large, and thus we judge that this area has a lower level of safety and should be warned as "danger zone."

When the average speed reaches the critical value (about 80 km/h), the standard deviation of the average speed reaches its maximum. It explains the number of vehicles at generally faster speed and slower speed both reaching a certain amount on the road. According to the research of Wu and Wu (2008), there is a positive correlation between the variable coefficient of speed and the incidence of traffic accidents. We can conclude that when vehicles are moving on the road at generally faster and slower speeds, and the number of vehicles at each speed are approximately the same, the road safety reaches the worst level. This finding lays a theoretical foundation for judging real-time road safety in a national road safety detection system and provides a new direction for road safety detection as well.

In order to make most use of the data and construct the detection indicators in relating to speed distribution, this chapter fits the speed on each road section by the maximum likelihood method. After fitting the distribution of samples, we judge the road safety through the characteristics of these distributions.

To fit the speed distribution of each road section, we observe possible nuclear density curves of each road section at first and pick out of several possible distribution curves, and then we calculate parameters of these possible distributions by the maximum likelihood function (Wang, 2009) and finally compare their AIC information to obtain the most suitable distribution. The smaller the number of AIC akaike information is, the better the fitting effect of distribution is. The results of nuclear density calculation are shown in Figure 11.3. We can see that speeds on a few road sections present unimodal distribution. Most road sections, however, generally present bimodal distribution. As there exist two kinds of distribution, we adopt normal distribution and multivariate normal distribution to fit the 20 road sections.

For the multivariate normal distribution (Mei, Qian, & Lin, 2003), when $k$ ($k \geq 2$) is the multivariate normal distribution, its probability density function is

$$p(\mathrm{x}) = \sum_{i=1}^{k} \pi_i p_i(x)$$

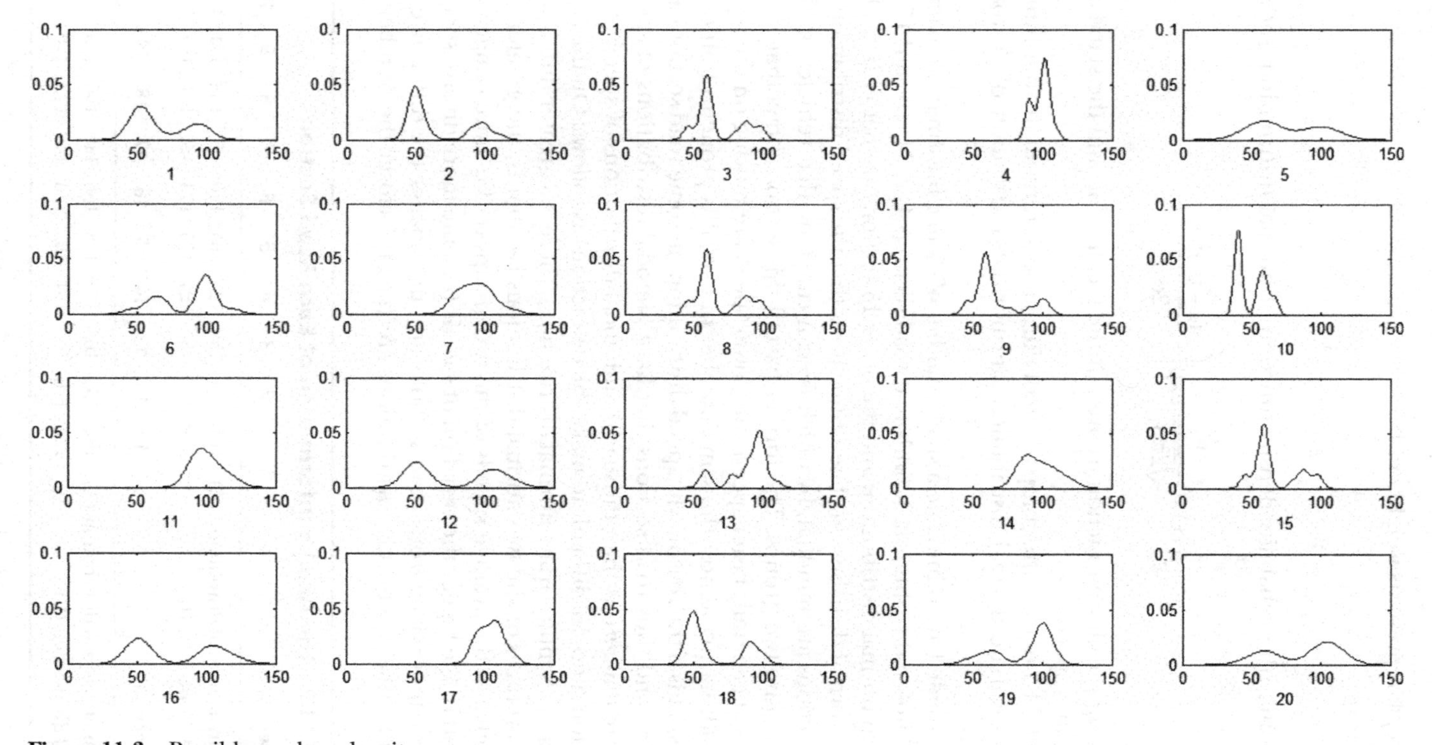


**Figure 11.3** Possible nuclear density curves.

Here, $\pi_i$ represents weight, and

$$\sum_{i=1}^{k} \pi_i = 1.$$

$p_i(x)$ is the $i$th probability density function of normal distribution, shown below:

$$p_i(x) = \frac{1}{\sqrt{2\pi\sigma_i}} \exp\left(-\frac{(x-\mu_i)^2}{2\sigma_i^2}\right)$$

Among the above function, represent the mean value and the standard deviation respectively.

Table 11.2 depicts the comparison of the akaike information after fitting the multivariate normal distribution and normal distribution of 20 road sections.

The shaded areas indicate a small number of akaike information, namely, the chosen distribution. It finds that road sections 4, 7, 11, 13, and 17 fit better in normal distribution, and the other 15 sections are better in multivariate normal distribution. The result shows that the speed distribution of most sections are bimodal, by which we can conclude that vehicles on the road fall into two groups. The former part of this section shows that when vehicles at generally faster speed and slower speed are moving on the road, with both having almost the same number of cars, it is prone to traffic accidents. This is the reason why speed distributions in some road sections are bimodal, and thus road sections that have bimodal distributions of speed are more dangerous than those with unimodal distributions of speed.

It should not be difficult to make the deductions as follows: On the road sections with multivariate normal distributions, there is a greater difference in the averages of the two normal distributions is—that is, the greater the difference of the average speeds of the two groups is, the more dangerous the road is; the bigger the variation of speeds (i.e., the more diverse speeds are), the more dangerous road sections are; the smaller the difference of weight for the two normal distributions are (i.e., the closer the number of

**TABLE 11.2 The Akaike Information of Each Road Section**

| Road section | 1 | 2 | 3 | 4 | 5 | 6 | 7 | 8 | 9 | 10 |
|---|---|---|---|---|---|---|---|---|---|---|
| Multivariate normal distribution (A) | 161 | 151 | 158 | 128 | 165 | 170 | 160 | 158 | 161 | 123 |
| Normal distribution (B) | 172 | 177 | 165 | 127 | 175 | 174 | 152 | 165 | 167 | 146 |
| **Road section** | **11** | **12** | **13** | **14** | **15** | **16** | **17** | **18** | **19** | **20** |
| Multivariate normal distribution (A) | 150 | 155 | 163 | 149 | 158 | 156 | 148 | 146 | 163 | 158 |
| Normal distribution (B) | 144 | 186 | 161 | 151 | 165 | 186 | 140 | 173 | 172 | 178 |

vehicles at faster speed and slower speed is), the more dangerous the road section is. Taking these three aspects that affect road safety into consideration, we construct a road safety indicator, the coefficient $W$, as follows:

$$W = \frac{|\mu_1 - \mu_2||\sigma|}{|\pi_1 - \pi_2|}$$

In this formulation, $\sigma$ represents the standard deviation of the multivariate normal distribution $\mu_1$, $\mu_2$ represents the average of two normal distributions, and $\pi_1$, $\pi_2$ represent weight of them respectively. $W$ increases as the increase of the standard deviation and the difference between speed averages; at the same time, W increases as the decrease of weight difference. The bigger the $W$ is, the more dangerous the road section is.

Table 11.3 shows the *W*s of 15 road sections with multivariate normal distribution. $W$ of section 12 is the biggest, so it is regarded as the most dangerous road section. By contrast, $W$ of section 14 is the smallest, so this section is the safest one.

We may draw a conclusion as follows: Traffic accidents occur more often on the roads with the coefficient of skewness than the roads without it. The coefficient of skewness can manifest the safety degree of the road. The bigger the coefficient of skewness is, the more easily traffic accidents occur. Here is the formulation for the coefficient of skewness:

$$\gamma_1 = E\left[\left(\frac{x-\mu}{\sigma}\right)^3\right]$$

We work out the coefficient of skewness of the 20 road sections by fitting distribution and sort them in descending order in Table 11.4. Road

**TABLE 11.3 *W*s of Each Road Section**

| Road section | 12 | 16 | 5 | 20 | 10 | 6 | 2 | 1 |
|---|---|---|---|---|---|---|---|---|
| W | 31117 | 30112 | 5635 | 3978 | 3641 | 3096 | 2899 | 2812 |
| Road section | 18 | 19 | 8 | 15 | 3 | 9 | 14 | |
| W | 2334 | 2154 | 1462 | 1462 | 1450 | 1170 | 1092 | |

**TABLE 11.4 The Coefficient of Skewness of Each Road Section**

| Road section | 9 | 2 | 19 | 18 | 8 | 15 | 3 | 1 | 6 | 20 |
|---|---|---|---|---|---|---|---|---|---|---|
| Coefficient of Skewness | 1.08 | 0.78 | 0.75 | 0.75 | 0.72 | 0.72 | 0.71 | 0.56 | 0.5 | 0.49 |
| Road section | 14 | 5 | 10 | 16 | 12 | 4 | 7 | 11 | 13 | 17 |
| Coefficient of Skewness | 0.32 | 0.31 | 0.24 | 0.15 | 0.15 | 0 | 0 | 0 | 0 | 0 |

section 9 has the biggest coefficient of skewness. It indicates that road safety is comparatively low and traffic accidents are prone to occur. Road section 12 has the smallest coefficient of skewness, so it is much safer.

### Construction of the Warning Indicators from the Dimension of Vehicles

Despite the real-time digits and distributions of speed, the speed change of vehicles on the whole highway is also one important aspect for road safety. In order to observe speed volatility on different road sections and discover dangerous road sections and vehicles, we sort vehicles according to their speeds on different road sections, then process outliers, and finally utilize local regression method to analyze changing speeds of different moving vehicles and propose the warning indicators based on the speed volatility. Among these fitting results of speed distribution on each road section, speed of most road sections presents multivariate normal distribution; thus, we conclude that the moving vehicles on these road sections fall into two groups. Clustering analysis is used to classify vehicles by calculating the number of Euclidean distance between speeds of different vehicles. We put vehicles with small number of ED into a group in turn. The calculating method of Euclidean distance between vehicle $i$ and vehicle $j$ is:

$$d = \sum_{k=1}^{20} (v_{ki} - v_{kj})^2$$

The clustering results are as follows: Group A includes vehicle 2, 3, 4, 5, 6, 7, 9, 11, 12, 14, 15, 16, 17, and Group B contains vehicle 1, 8, 10, 13, 18, 19.

As the actual data collection is usually not completes, the recognition and process of outliers become indispensable steps for data application. "The outlier is a data point that is very much bigger or smaller than the next nearest data point" (Wang & Tong, 2006, p. 67). In our research, outliers are the speed data which differ greatly from that of other vehicles on one or several certain road sections. In Figure 11.4, although they belong to Group A, vehicle 2 and vehicle 4 have a larger number of Euclidean Distance than other vehicles, which makes them the possible outliers.

We use the *LOF* method to recognize outliers. By measuring the mean value of the available distance between data points and their nearby points, the method judges whether a data point is greatly deviant to other data. The specific steps which is described by (Markus, 2000) are as follows:

1. For each data point $p$, when calculating its $K$ distance, $K-distance(p)$, it must satisfy:
   I. There exist at least $K$ objects to make $d(p, q) \leq d(p, o)$
   II. There exist at most $K-1$ objects to make $d(p, q) < d(p, o)$

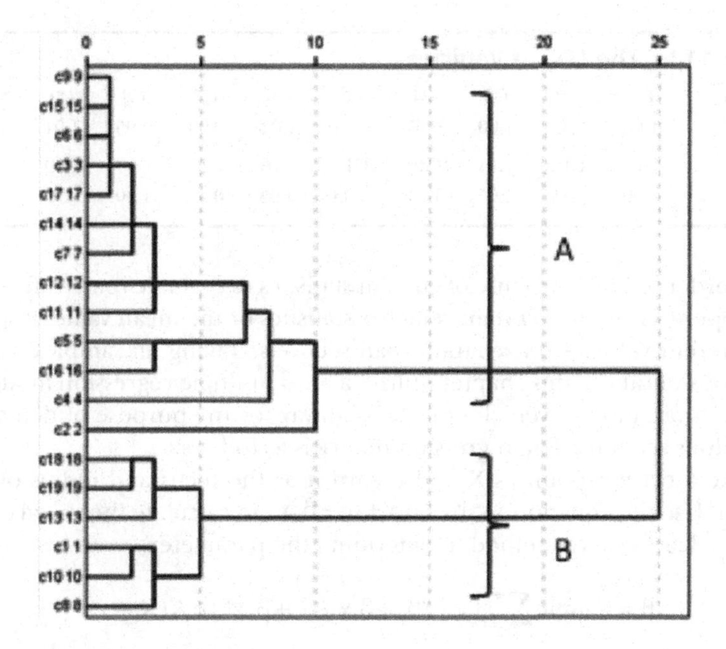


**Figure 11.4** Clustering for vehicles.

$d(p, o)$ is defined as $K-distance(p)$, and the neighborhood of $p$ is called neighborhood K, which takes k–distance as the radius.

2. Calculating the local available density of object $p$

$$Ird_{M\ in\ pts}(p) = \frac{|N_k(p)|}{\Sigma_{o\ ink} reach_{dist_k}(p,o)}$$

$|N_k(p)|$ presents the number of points $reach_{dist_k}(p,o) = max(K-distance(p), d(p,o))$in the neighborhood $k$.

3. Calculating $LOF(p)$:

$$LOF(p) = \frac{\Sigma_{o \text{ in } k} \frac{Ird_k(o)}{Ird_k(p)}}{|N_{k-\text{nearest}}(p)|}$$

If $LOF(p)$ of one point is close to 1, then it is not an outlier; otherwise, the more deviant to 1 one point is, the more possible it is an outlier. It selects $N$ points as the outlier, which have larger number of $LOF(p)$.

When $k$ is 5, we adopt the method of Euclidean distance to measure the distance, and then we obtain $LOF$ of 19 vehicles respectively, as shown in Table 11.5. From Table 11.5 we can find that outliers are vehicles 2, 4, 5, and 16. We reject these four vehicles in the following processing for easy operations.

**TABLE 11.5 The *LOF* of Vehicles**

| Vehicle | C1 | C2 | C3 | C4 | C5 | C6 | C7 | C8 | C9 | C10 |
|---|---|---|---|---|---|---|---|---|---|---|
| | 1.03 | 6.99 | 1.00 | 6.87 | 6.94 | 1.01 | 0.99 | 0.95 | 1.01 | 1.02 |
| Vehicle | C11 | C12 | C13 | C14 | C15 | C16 | C17 | C18 | C19 | |
| | 0.99 | 1.14 | 1.05 | 1.31 | 0.98 | 5.84 | 0.95 | 1.00 | 0.95 | |

In order to fully make use of current statistics to evaluate roads safety, it is improper to refer to a certain vehicle's statistics or the mean value of speed of different vehicles on separate road sections. Taking all samples' speed into consideration, this chapter utilizes a local multiple regression model to fit speed variation of vehicles on the highway for the purpose of detecting hazardous sections. The regression theory is as follows:

Take a certain point as X and regard it as the focus and radius of the region. Use the equation $y = \beta_0 + \beta_1 x + \cdots + \beta_n x^n$ to calculate the speed curve and the least square method to determine the parameter.

$$\beta = \operatorname{argmin} \sum_{i=1}^{n} \left( y_i - (\beta_0 + \beta_1 x + \cdots + \beta_n x^n) \right)^2 K(X_i - x)$$

$K(X_i - x)$ is the weighting function, and $\hat{y}(x) = \hat{\beta}_0 + \hat{\beta}_1 x + \cdots + \hat{\beta}_n x^n$ is the estimated number.

The statistics of the chapter are the average rather than the instant speed of vehicles, and the chapter supposes that vehicles on the section conduct the regression based on the average speed. The result is Figure 11.5. The regression result depicts that there is a distinct disparity between two groups, which corresponds to the bimodal distribution of speed.

When moving on the highway, if speed changes dramatically in adjacent sections, vehicles are prone to traffic accidents, for example, rear-end collision. *Highway Route Design Specification* promulgated by the Ministry of Communication in 1994 explicitly points out: "Speed differences of adjacent road sections on a highway within the scope of one terrain partition

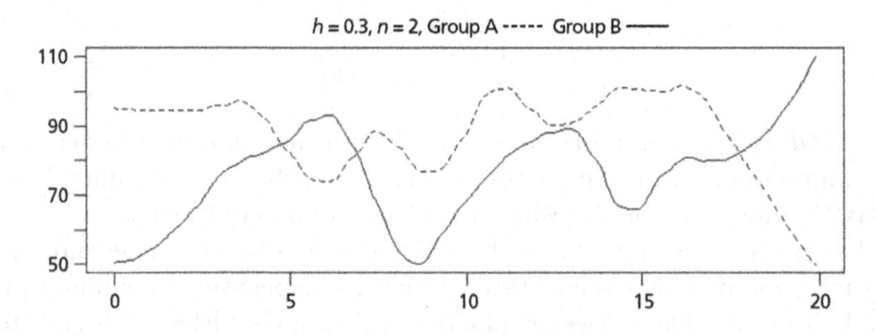


**Figure 11.5** The speed binomial regression of Group A and Group B.

should not exceed 20 km/h" (Ministry of Transportation, Ministry of Communications, 1994, p. 4).Therefore, when speed variance of a vehicle in two adjacent sections exceeds 20 km/h, there exists potential safety risk and the early warnings need to be issued.

We can easily find that there are great speed variances of the two groups in adjacent sections. Speeds of Group B are much faster between adjacent sections with less fluctuation. Speed of Group A has great fluctuation in adjacent sections. It exceeds 20 km/h in adjacent sections from section 7 to 8, and from 8 to 9, for which a safety risk exists to a great extent. At the same time, because of the greater speed volatility of Group A than Group B, it is judged that risk factors of potential safety hazard exist in Group A, and safety warning need to be issued.

## CONCLUSION REMARKS

In this chapter, we have focused on the road safety detection in relating to real-time environmental speed and have developed a road safety detection system based on real-time environmental speed data that are collected from vehicles moving on the highway from Beijing to Beidaihe. We have investigated three aspects of information including variance and average speed, speed distribution, and speed volatility on two dimensions of vehicle and speed. A road safety warning system was developed through several real-time indicators that includes the coefficient of variation, skewness, and speed variance of adjacent sections. We have provided a practical reference for road safety detection and a new direction of road safety researches. There are still some limitations in application of our developed system due to the limited data, and further research of verifications need to be done.

## ACKNOWLEDGMENT

Work was partially supported by grant project No13XNI011 from Center of Applied Statistics, Ministry of Education of the People's Republic of China. The financial support from National Natural Science Foundation of China (Grant # 71110107024) is acknowledged. We also appreciate Chunling Wu, Qile Yang, and Yan Zhong for part computing and discussions.

## REFERENCES

Feng, Z., & Liu, J. (2008). Construction of road traffic warning system. In National Steering Group on Intelligent Transport Systems Coordination (Ed.), *The*

*Fourth China Intelligent Transportation Annual Conference Proceedings* (Vol. 6, pp. 59–62). Shenzen, China: Author.

Ministry of Transportation, Ministry of Communications. (1994). *Highway route design specifications.* First Highway Survey and Design Institute, 7: 4–10. Beijing, China: Author.

Ma, D., Liu D., & Zheng Y. (2009). Speed on road traffic safety and the countermeasures. *Chinese People's Public Security University* (*Natural Science*), *61*, 59-62

Breunig, M., Kriegel, H.-P., Ng, R. T., & Sander, J. (2000). LOF: Identifying density-based local outliers. In *Proceedings of the 2000 ACM SIGMOD international conference on management of data,* pp. 93-104, doi:10.1145/335191.335388

Mei, G., Qian, Q., & Lin, D. (2003). Highway bridge vehicle loads bimodal distribution probability model. *Tsinghua University* (*Natural Science*), *10,* 1394–1396, 1404.

Shao, Z. (2005). Urban road traffic safety design and application of early warning indicators. *Police Technology, 1,* 39–41.

Wang, H., & Tong, Y. (2006). Progress outlier mining. *Intelligent Systems, 1,* 67–73.

Wang, X. (2009). *Nonparametric statistics.* Beijing, China: Tsinghua Publishing.

Wang, X., & Liu, D. (2010). Early warning indicator system of urban road traffic safety. *Chinese People's Public Security University* (*Natural Science*), *2,* 58–61.

Wu, Y., & Wu, Z. (2008). Road safety analysis based on the average speed and the speed standard deviation method. *Highways Technology,* 25(3), 139–142.

# ABOUT THE EDITORS

**Kenneth D. Lawrence** is a Professor of Management Science and Business Analytics in the School of Management at the New Jersey Institute of Technology. Professor Lawrence's research is in the areas of applied management science, data mining, forecasting, and multi-criteria decision-making. His current research works include multi-criteria mathematical programming models for productivity analysis, discriminant analysis, portfolio modeling, quantitative finance, and forecasting/data mining. He is a full member of the Graduate Doctoral Faculty of Management at Rutgers, The State University of New Jersey in the Department of Management Science and Information Systems and a Research Fellow in the Center for Supply Chain Management in the Rutgers Business School. His research work has been cited over 1,500 times in over 225 journals, including: *Computers and Operations Research, International Journal of Forecasting, Journal of Marketing, Sloan Management Review, Management Science,* and *Technometrics.* He has 301 publications in 28 journals including: *European Journal of Operational Research, Computers and Operations Research, Operational Research Quarterly, International Journal of Forecasting* and *Technometrics.* Professor Lawrence is Associated Editor of the *International Journal of Strategic Decision Making* (IGI Publishing). He is, also, Associated Editor of the *Review of Quantitative Finance and Accounting* (Springer Verlag), as well as Associate Editor of the *Journal of Statistical Computation and Simulation* (Taylor and Francis). He is Editor of *Advances in Business and Management Forecasting* (Emerald Press), Editor of *Applications of Management Science* (Emerald Press), and Editor of *Advances in Mathematical Programming and Financial Planning* (Emerald Press.)

*Contemporary Perspectives in Data Mining, Volume 2,* pages 229–230

**Ronald K. Klimberg, PhD** is a Professor in the Department of Decision and System Sciences of the Haub School of Business at Saint Joseph's University. Dr. Klimberg has published 3 books, including his recent book *Fundamentals of Predictive Analytics using JMP,* edited 9 books, over 50 articles and has made over 70 presentations at national and international conferences. His current major interests include multiple criteria decision making (MCDM), multiple objective linear programming (MOLP), data envelopment analysis (DEA), facility location, data visualization, data mining, risk analysis, workforce scheduling, and modeling in general. He is currently a member of INFORMS, DSI, and MCDM. Ron was the 2007 recipient of the Tengelmann Award for his excellence in scholarship, teaching, and research.

Printed in the United States
By Bookmasters